Electronic Waste

**Recent Titles in the
CONTEMPORARY WORLD ISSUES
Series**

Rape and Sexual Assault
Alison E. Hatch

Campus Free Speech
Lori Cox Han and Jerry Price

Global Terrorism
Steven J. Childs

Food Insecurity
William D. Schanbacher and Whitney Fung Uy

Conspiracy Theories
Jeffrey B. Webb

Advertising in America
Danielle Sarver Coombs

Cyber Warfare
Paul J. Springer

Extremism in the Police
Carla Lewandowski and Jeff Bumgarner

The U.S. Criminal Justice System
Sarah Koon-Magnin and Ryan J. Williams

Women and Girls in STEM Fields
Heather Burns Page

Remote Learning and Distance Education
William H. Pruden III
Religious Belief and Science
Glenn H. Utter

Child Welfare in America
Yvonne Vissing

Water in the West
Jacqueline Vaughn

Teen Runaways in America
Michele Wakin

Electronic Waste

A Reference Handbook

Josh Lepawsky

BLOOMSBURY ACADEMIC

NEW YORK · LONDON · OXFORD · NEW DELHI · SYDNEY

BLOOMSBURY ACADEMIC
Bloomsbury Publishing Inc
1385 Broadway, New York, NY 10018, USA
50 Bedford Square, London, WC1B 3DP, UK
29 Earlsfort Terrace, Dublin 2, Ireland

BLOOMSBURY, BLOOMSBURY ACADEMIC and the Diana logo are
trademarks of Bloomsbury Publishing Plc

First published in the United States of America 2025

Copyright © Bloomsbury Publishing Inc, 2025

Cover image: Fabrizio Troiani/Alamy Stock Photo

All rights reserved. No part of this publication may be reproduced or transmitted
in any form or by any means, electronic or mechanical, including photocopying,
recording, or any information storage or retrieval system, without prior
permission in writing from the publishers.

Bloomsbury Publishing Inc does not have any control over, or responsibility for, any
third-party websites referred to or in this book. All internet addresses given in this
book were correct at the time of going to press. The author and publisher regret
any inconvenience caused if addresses have changed or sites have ceased
to exist, but can accept no responsibility for any such changes.

A catalog record for this book is available from the Library of Congress.

ISBN: HB: 979-8-2161-7013-6
         ePDF: 979-8-2161-7014-3
         eBook: 979-8-2161-7015-0

Series: Contemporary World Issues

Typeset by Integra Software Services Pvt. Ltd.
Printed and bound in the United States of America

To find out more about our authors and books visit www.bloomsbury.com
and sign up for our newsletters.

Contents

Preface		ix
Acknowledgments		xiii
1	**Background and History**	**1**
	From Modern Waste to E-waste	2
	Strategies for Thinking Critically about E-waste	8
	E-waste: A Global Synopsis	11
	Estimates of Total Mass and Growth Rates of E-waste Arising	11
	Transboundary Shipments: Patterns and Processes	15
	Wastes Are Not, They Are Made	19
	E-waste Regulation around the World	21
	The Basel Convention	22
	The Bamako Convention	24
	Africa and Middle East/West Asia	25
	Americas	26
	Asia Pacific	27
	Europe	31
	Conclusion	32
	References	33
2	**Problems, Controversies, and Solutions**	**39**
	The End-of-Pipe Problem	39
	Classification and Regulatory Definitions of E-waste	45
	Magnitude: How Big a Problem Is E-waste?	48
	Scale: Matching Problems with Solutions	51
	Jevons Paradox, Efficiency, and Rebound	52
	Decoupling	55
	Transboundary Flows	57
	(Il)legality	60
	E-waste, Environmental Racism, Environmental Justice, and Colonialism	62
	Carbon Fixation	68

Solutions: The Waste Hierarchy	71
Disposal	72
Recycling	74
Reuse	78
Reduce, Avoid, Prevent	81
Design Solutions	86
Reorganizing for Solutions	89
Conclusion	93
References	94

3	Perspectives	103
	The Light at the End-of-Life of Electronic Equipment: Narratives from the Global South *Ramzy Kahhat*	105
	Wasted Definitions: Negotiating Truth in Environmental Journalism *Adam Minter*	108
	A Closer Look at Agbogbloshie and the Electronic Waste Narrative *Grace Abena Akese*	112
	Welcome to Solution? Repurposing of Electronic Waste in Dar es Salaam, Tanzania *Samwel Moses Ntapanta*	117
	Designing for Reuse *Melissa Gregg*	121

4	Profiles	127
	Agbogbloshie Maker Space	127
	American Chamber of Commerce in Europe	128
	Bamako Convention	129
	Basel Action Network	130
	Basel Convention	131
	Bureau of International Recycling	133
	Center for Public Environmental Oversight	133
	Circular Electronics Partnership	134
	Commission for Environmental Cooperation	135
	Chatham House	136
	DigitalEurope	137
	Electronic Products Recycling Association	138
	Electronics Watch	138
	E-Scrap News	139
	E-Stewards	140

Eurostat	140
FreeGeek	142
Global Electronics Council	143
Global Enabling Sustainability Initiative	144
Global E-waste Statistics Partnership	144
Good Electronics Network	145
Greenpeace	146
iFixit	147
Information Technology Industry Council	147
International Campaign for Responsible Technology	148
International Solid Waste Association	149
International Telecommunications Union	150
Interpol	150
Oeko-Institut	151
Open Repair Alliance	152
Organisation for Economic Co-operation and Development	153
reBOOT	153
Recycled Materials Association	154
Recycling Industry Operating Standard	154
Repair Association	155
Repair Café Movement	156
Repair Europe	156
Responsible Business Alliance	157
Restart Project	158
Santa Clara Center for Occupational Safety and Health	159
Silicon Valley Toxics Coalition	160
Solve the E-waste Problem	160
Sustainable Electronics Recycling International	161
United Nations Conference on Trade and Development	162
United Nations Environment Programme	163
United Nations Industrial Development Organization	163
United Nations Institute for Training and Research	164
United Nations University	164
United States Federal Trade Commission	165
United States Public Interest Research Group	166
WEEE Forum	166
WEEELABEX	167

viii   *Contents*

World Computer Exchange	168
World Health Organization	168
World Reuse, Repair, and Recycling Association	169
WorldLoop	170

5   Documents   173

"Recovery of Precious Metals from Electronic Scrap,"
    United States Department of the Interior (1972)   173

"Flows of Selected Materials Associated with World Copper
    Smelting," United States Geological Survey (2005)   174

"Fairchild, Intel, and Raytheon Sites Middlefield/Ellis/Whisman
    (MEW) Study Area Mountain View, California, Record of
    Decision," United States Environmental Protection Agency (1989)   175

"2011–2017 Greenhouse Gas Reporting Program Industrial Profile:
    Electronics Manufacturing Sector," United States Environmental
    Protection Agency (2018)   179

"Center for Corporate Climate Leadership Sector
    Spotlight: Electronics," United States Environmental Protection
    Agency (2016)   182

"Countering WEEE Illegal Trade (CWIT) Summary
    Report, Market Assessment, Legal Analysis, Crime Analysis and
    Recommendations Roadmap" (2015)   186

"Future E-waste Scenarios," United Nations University/
    United Nations Environment Programme (2019)   190

"Nixing the Fix: An FTC Report to Congress on Repair Restrictions,"
    United States Federal Trade Commission (2021)   198

6   Resources   203

Books	203
Articles and Reports	209
Journals	224

7   Chronology   231

Glossary	239
Index	244
About the Author	253

Preface

Electronics are increasingly common devices in more and more peoples' lives. Smartphones, laptops, tablets, and other forms of personal computers are important parts of how people organize and carry out their day-to-day activities in relationships of work, play, friends and family, and more. As electronics become just so much ordinary infrastructure, they also bring with them all sorts of consequences, including environmental impacts. This book is about the pollution and waste that arise from electronics.

Sometime in the decade between 2030 and 2040 people around the world will discard approximately 100 million metric tons of electronics annually if trends that are current today continue unchanged. One hundred million tons of electronics discarded by consumers may sound like a big problem, but today's aggregate pollution and waste arising from the mining for, and manufacturing of, electronics already exceed these future projections. For example, the electronics manufacturing sector is one of the largest industrial consumers of copper. Today, a single copper mine generates as much waste per year as what is estimated will arise from post-consumer discards globally between 2030 and 2040. Meanwhile, just the manufacturing of screens for electronic devices is estimated to have released the equivalent of 10–15 million tons of carbon dioxide in 2020 and electronics manufacturing facilities located in only three countries—Canada, Mexico, and the United States—collectively released about 447,000 metric tons of hazardous chemicals between 2006 and 2020. More recent data show that electronics manufacturing facilities in the United States alone release more than 39,000 tons of hazardous polyfluoroalkylated compounds (PFAS, or "forever chemicals") per year.

All these numbers may be a little overwhelming, but they point to two overarching issues with respect to pollution and waste arising from electronics: if pollution and waste from mining and manufacturing for electronics are accounted for, then the magnitude of future estimates of mass of discarded electronics is already here. Also, pollution and waste arise ubiquitously, but unevenly, throughout the full lifecycle of electronics, not just when consumers get rid of their devices.

x *Preface*

Mitigating and eliminating the negative environmental consequences of electronics across their lifecycle are major challenges. *Electronic Waste: A Reference Handbook* is a practical guide for students and general readers who wish to understand the pollution and waste implications of electronics from the mining for the necessary materials, from their manufacture, from their use, and at their end-of-life when discarded. The book uses examples from various locations around the world to cover important background and historical information that helps readers place current concerns about the environmental implications of electronics in context. It also provides an overview of key problems, controversies, and possible solutions to the issues of pollution and waste arising from electronics. These discussions are complemented by profiles of key actors related to these issues, primary document sources, and resources for further learning and research about these topics.

Electronic Waste: A Reference Handbook is organized into seven chapters and includes a glossary of key terms and an index. Chapter 1, "Background and History," situates current concerns about managing electronic waste (e-waste) as a post-consumer waste management problem within broader patterns and processes that have led to contemporary forms of pollution and waste that arise from the industrial production of electronic devices. The chapter pays attention to both conceptual and empirical developments around pollution and waste issues associated with electronics. It discusses strategies for thinking critically about e-waste and provides a global synopsis of the problem including an overview of relevant regulation using examples from around the world.

Chapter 2, "Problems, Controversies, and Solutions," delves into several enduring problems and controversies associated with e-waste, including the very definition of the issue. Among the topics discussed is that of problem framing and how such framing plays into the possibilities for imagining and implementing various solutions. One of the key problems associated with e-waste is that much of the contemporary discussion presupposes that such a waste emerges only after consumers get rid of devices. Yet, if solutions are to be found to overall pollution and waste arising from electronics throughout their lifecycle, then different ways of framing the problem are needed as are different solutions needed beyond consumer-based repair and recycling on their own.

Chapter 3, "Perspectives," comprises five short essays by authors written from their various personal and professional points of view. These essays include authors with backgrounds in academia, industry, and journalism. Each of the authors engages deeply with different aspects of the e-waste problem and brings rich empirical knowledge and experience of the topic that speaks to the broader issues addressed in Chapters 1 and 2.

Chapter 4, "Profiles," offers an alphabetized list and brief description of key actors focused on the issues of pollution and waste arising from electronics. Some of these actors are governments and governmental bodies, others are environmental nongovernmental organizations (ENGOs), key international legal agreements, and trade associations, among others. The chapter describes the roles that each individual actor plays in shaping the public discourse on e-waste.

Chapter 5, "Documents," provides excerpts from primary documents relevant to understanding pollution and waste arising across the lifecycle of electronics. Some of these documents are governmental reports while others are research documents from intergovernmental organizations, such as the United Nations.

Chapter 6, "Resources," offers an annotated list of key books, articles, and reports that cover issues of pollution and waste arising from the mining, manufacturing, and use of electronics as well as how repair and recycling fit into the broader picture of pollution mitigation.

Chapter 7, "Chronology," provides a timeline of key events leading up to the contemporary concern about e-waste as a post-consumer waste management problem. In so doing, the chapter tracks the evolution of the problem from being associated with the exposure of workers to toxicants on the factory floor to, much more recently, a narrower framing of the problem as one of post-consumer waste management.

A glossary of key terms can be found following Chapter 7. The book concludes with an index to help the reader navigate the text.

Acknowledgments

This book would not exist were it not for an invitation to write it from Kevin Hillstrom at Bloomsbury. I am grateful for the support and assistance Kevin offered me during the crafting of the manuscript. My sincere thanks also to development editor Nicole Azze whose careful attention to the manuscript helped me hone it to a crisper and clearer text.

A special thanks to the authors whose essays comprise "Chapter 3: Perspectives" of this book. I deeply appreciate your willingness to contribute to this project and thereby substantially enrich it.

To my colleagues at Memorial University and the Department of Geography, in particular, thank you for making our university a meaningful place worth fighting for.

Erin and Arlo: I love you. You're the best to hang out with.

Alright Sorrel, come on … let's go for a walk.

1

Background and History

The term "electronic waste" (e-waste) usually refers to electrical and electronic equipment that is discarded by consumers. In this sense, e-waste is understood as an especially problematic fraction of overall waste arising because of the risks it poses for toxic pollution and for the loss of potentially reusable devices and valuable materials. Tens of millions of tons of post-consumer e-waste are estimated to arise globally every year and at a growing rate. Current estimates suggest 62,000,000 metric tons (62 megatons or Mt) of e-waste arose in 2022 (Baldé et al. 2024) and that more than 100 Mt could arise by 2030. Concerns about post-consumer e-waste began to emerge in the early 2000s. However, those earlier concerns have more recently expanded to also include pollution and waste arising before consumers get rid of their devices. This more recent and expanded understanding of e-waste also looks to the pollution and waste implications of the mining for and manufacturing of electronics. This chapter provides an overview of the history and current approaches toward measuring and managing e-waste as a post-consumer waste management problem, including key regulatory frameworks from examples around the world. It also situates current management approaches to e-waste within relevant wider discussions of the history and meaning of waste and how the line between waste and resource is drawn.

Chuquicamata copper mine in Calama, Chile, one of the biggest open pit copper mines in the world. The electronics sector is the second largest consumer of global copper resources after the building and construction industry. Estimates show that a 1% reduction in annual mining waste at Chuquicamata would equate to a 28% reduction in annual waste at the end-of-pipe consumer discard phase in the European Union. (Nicola Messana/Dreamstime.com)

From Modern Waste to E-waste

To better understand e-waste, it is helpful to examine some key historical and geographical characteristics that make up the broader category of what has been called "modern waste." According to Samantha MacBride, a professor and waste management professional, modern wastes share the characteristics of being "synthetic, unpredictable, and above all heterogenous" (MacBride 2012, 174). Plastics are a quintessential example of modern waste. Mass production of plastics largely begins after the Second World War. In that sense they are a very recent introduction to the planet's economic and biological systems. They are also made of chemical mixtures that do not occur in nature without the intervention of scientific knowledge and industrial technology. In this sense plastics are synthetic. They are also unpredictable in large part because their chemical mixtures are proprietary. That means that no one other than small groups of industrial chemists employed by chemical manufacturing firms knows what the actual underlying chemical constituents are. Downstream users of those plastics (such as consumers) and processors (such as recyclers) have no way of knowing for sure what particular mixture of chemical compounds this or that plastic they are using or handling is made of. That inability to know has important consequences, especially for processors who may wish to try to return recycled plastics back into manufactured products. Mixing of different plastics often occurs in processing and results in degraded quality, making the recycled plastics less durable and having to be "downcycled." Meanwhile, not only are plastics themselves made up of proprietary mixtures of chemicals and are in that sense heterogeneous, but plastics also might be incorporated into products that are themselves made of multiple synthetic materials. Electronics are a good example of this kind of heterogeneity. A typical smartphone, laptop, or desktop computer is usually comprised of mixtures of plastics, metals, and glasses. It is important to understand each of these material types in the plural. Even the same laptop will incorporate several different types of plastics, for example, those that make up the external housing, wire coatings, and those on which microchips and other components might be mounted. Many different types of metals are also incorporated into electronics, such as aluminum, copper, steel, neodymium, cobalt, and tantalum, among others. Even a smartphone display, which users might experience as a single thing, is made up of different plastics, metals, and glasses. Electronics discarded as "e-waste" definitely meet MacBride's meaning of modern waste.

Background and History

Waste is a word used in everyday language and what it is might seem self-evident. In fact, waste is quite difficult to define. Delving a little bit into the etymology of the word might give some indication of why waste defies easy definition. The Anglophone word "waste" comes from the Latin word *vastus*. *Vastus* means unoccupied, uncultivated, void, and immense. *Vastus* also has etymological links to Sanskrit that means wanting or deficient. Used in everyday conversations today the word "waste" can take on many different meanings depending on context. For example, it can refer to something that is discarded, unwanted, unusable, or a by-product of some process. But it can also suggest being frivolous or needlessly profligate as when someone is said to be wasting time.

There may be no way to define waste in some universally applicable sense. Consequently, it can be helpful to think of waste as always being situated within some specific context or other. One way to illustrate the situated character of waste is to recall that all life sustained by oxygen, including human life, depends for its existence on the continual renewal of that gas which is a metabolic by-product (i.e., waste) of photosynthesis (Volk 2004). Humans and other oxygen-dependent lifeforms in a certain practical sense dwell at the bottom of an atmospheric ocean of waste gas discarded by other life without which they would perish. At the same time, this example of the situatedness of waste is somewhat misleading because waste is also infused with a wide range of values. Some of those values can be economic—for example, when a waste hauling company collects material outside an office building from a disposal bin and charges businesses in that building for those services. But waste can also connote a broader meaning of values beyond the economic including those relating to moral, ethical, even spiritual values. Think of the moral judgment implicit in the idea of wasting time. Waste can have political connotations as well. For example, spending public money on services such as healthcare or education might be deemed wasteful by some if the money is spent on people, places, or things deemed undeserving. In these broader ways, waste has a cosmological significance; that is, what gets defined as waste is indicative of the worldview of the people who get to define what will count as waste and what will not. Waste provokes philosophical thought and concrete measures to avoid, disperse, dilute, and rework—in short, manage—it.

Beyond its etymological roots, the modern Anglophone meaning of "waste" has its origins in the long history of enclosure that turned idle land not being used for cultivation—that is, waste lands—into enclosed private property.

Wastes in the form of common land entered the long history of legal modernity with the Magna Carta (Gidwani 2012). The Great Charter of the Forest of 1225 defended common rights to common land—known as wastes—but soon after with the Statute of Merton of 1235 those rights were patchily and unevenly eroded. That erosion was a lengthy process, covering at least 200 years. Along with the drama of monarchic attempts to regulate enclosure of waste lands by the manorial elite came efforts to mitigate the ill effects on those who were displaced, resulting in the emergence of institutions of the modern state apparatus and capital, such as courts devoted to legal enforcement of private property law. What emerged was a shift from wastes as commons to commons as wastes (Goldstein 2013). Enabling that shift were all the techniques of regulation and enforcement that depopulated the commons as wastes, often violently. In 1700 most land in England was still open waste land, but by 1840 at least 20 percent of it had been enclosed by acts of Parliament between 1750 and 1820. These enclosures chart a history of waste in its multiple, ramifying forms: initially it has a largely benign or even positive connotation as land available for common use, but it later comes to have strongly negative connotations. By the seventeenth century waste came to be understood as antithetical to the political society imagined by philosophers such as John Locke and Adam Smith in their notions of private property and associated ideas.

E-waste is tangled up with these long histories and wider meanings of waste and commons (Lane 2011). One might stop for a moment and ask why electronics have been singled out as a particularly problematic source of waste given the huge range of commodities that are part of many people's daily lives? Why today is there a seemingly identifiable and urgent problem called "e-waste" but not a similarly labeled problem called "c-waste" (car waste)? Actually, an historical lens on public concerns about "automobile graveyards" (Zimring 2011, 528) offers some answers to these questions. Over a period of roughly forty years between 1920 and 1960 in the United States, membership in automobile associations grew from a wealthy few to a broader upper- and middle-class car-owning cohort, especially after the Second World War. According to historian Susan Strasser (1999), Great Depression-era photography in the United States often used abandoned automobiles as shorthands for the broader economic crisis. Later, the Second World War massively increased manufacturing capacity in the United States, and that growth in capacity laid a foundation for mass production for, and consumption by, an expanded middle class (even as it was largely racialized as white). Automobile scrap heaps in pockets and fringes of expanding suburban

Background and History 5

landscapes clashed with the aesthetic ideals of the burgeoning, largely white middle class moving to those suburbs. And, by the 1970s, images of abandoned automobiles in the United States once again became symbolic stand-ins for broader concerns about environmental degradation, again expressed by a largely white middle class.

The linking up of specific commodities, such as cars, with more abstract aspirations (e.g., personal automobility as synonymous with freedom) and dread (e.g., scrap cars as indications of economic and environmental decline) offers important insight about similar factors in play regarding e-waste today. Arguably, an analogy can be made between historical public concern about scrap cars and contemporary concern about e-waste. Like cars before them, electronics have been infused with symbolic and allegorical meaning. As such, electronics have come to be important symbols that help tell stories with moral and political meanings beyond the brute materials of which electronics are composed. For example, for some critics, Apple's AirPods have come to symbolize forms of wasteful expenditures of money, materials, and energy (Haskins, Koebler, and Maiberg 2019; Taffel 2022). In stories like these, the wastefulness symbolized by a specific kind of electronic device is as much about moral values as it is about frivolous expenditures of money or thoughtless losses of valuable materials. Packed into electronics, then, are not just sophisticated components, but also dreams about hoped-for technological futures and nightmares about social and ecological damage to the planet.

Another side of the allegorical character of e-waste is its connections to disposability and planned obsolescence. Both disposability and planned obsolescence are perennial subjects of criticisms of mass consumption. Here too, a historical approach offers important insights into the character of debates about e-waste today. Disposability of products is not inevitable. Indeed, the very idea that products purchased for personal consumption should be disposable is a relatively new invention. For example, in the 1860s in the United States, paper shirt collars were marketed to men as a convenience, but they were also marketed to women, not for their own use, but as a labor-saving device and source of "freedom" from gendered divisions of labor around family and household reproduction (Strasser 1999). A broader shift away from individual and household relationships of stewardship of things to the disposability of things would not come to pass in the United States until after the Second World War.

As mentioned, one of the domestic consequences of the war was a massive buildup of industrial manufacturing capacity. That capacity enabled a broad economic shift toward mass production and consumption in the postwar period. Disposability became a way for manufacturers to overcome the potential threats to profits that such enhanced manufacturing capacity enabled. Consequently, instead of enticing consumers to buy things to put *in* their homes, manufacturers and advertisers encouraged consumers to buy things to move *through* their homes (Packard 1960). Yet, the transition to mass consumption and disposability was not smooth. Strasser (1999) recounts how soldiers actually rioted when presented with disposable paper cups for individual use in place of their traditional communal tin cup. Such a strong reaction to disposability is suggestive of a broader theme: people are not always and everywhere wasteful; waste and wastefulness arise with changes to infrastructure (e.g., paper cups substituted for a communal tin one) rather than from an essential "human nature" on its own.

Planned obsolescence offers manufacturers another solution to the problem of declining profits. Yet, like disposability, planned obsolescence is neither inevitable nor does it escape resistance. Interestingly, planned obsolescence was first put forward as a potential government-run solution to the Great Depression of the 1930s. To solve what some people understood as a crisis of over accumulation, the government should put a legally mandated "death date" on all products. Once the death date was reached products would be destroyed and consumers would purchase new ones, thus ensuring new rounds of exchange (London 1932). Only slowly were manufacturers and advertisers able to combine disposability and planned obsolescence into makeshift strategies to enhance their bottom lines. Like disposability, planned obsolescence was (and is) actively resisted. Indeed, in the pages of industry trade journals, some industrial engineers quite explicitly criticized planned obsolescence as a moral affront to their dignity and their broader societal responsibilities (Slade 2006).

As mentioned, these brief glances at the histories of disposability and planned obsolescence in the postwar United States demonstrate an issue of broader import: people are not inherently wasteful. There are historical and ongoing examples of resistance to disposability and planned obsolescence, even in societies in which mass production and consumption have become ordinary. Yet, if it can be said that people are not inherently wasteful, one might also rightfully ask: why is there so much waste? Again, answers can be found from the histories briefly discussed. In short, and as stated, waste is much more about infrastructure and much less about any essential wastefulness inherent in people's behavior (Liboiron 2014). Devices such as electronics can end up

Background and History 7

facilitating a reorganization of everyday activities that cannot easily be undone once that change in organization has taken hold. For example, in 2010 a *New York Times* journalist commenting on the different ways cellphones were being used in India compared to the United States was "struck by how often calls drop here [in the US] and surprised that text-messaging, so vital to Indians, has yet to entrench itself in America" (Giridharadas 2010, 1). The situation in the United States has changed remarkably in the last decade and a half. Texting to micro-plan around ordinary, everyday activities has become a norm in a country (the United States) where not so long ago it was an exception. In some sense, texting on a smartphone has moved from being a practice of upper echelons of business executives to being part and parcel of the infrastructure of organizing everyday routines for more and more people (Gregson 2023). Now, in 2024, it would be hard for a lot of people in the United States to get through an ordinary day without relying on texting to coordinate some activities. Of course, this is not true for every single person. Nor would it necessarily be *impossible* to coordinate one's everyday activities without texting, but it would be harder. The broader point here is that certain kinds of infrastructure—cellphones and their networks for example—have been widely incorporated into how everyday people organize daily routines. Cellphones are increasingly a piece of infrastructure that one cannot do without if one is to participate in the broader social ordering of everyday life. In part, what this means is that suggesting that someone simply give up their cellphone if they are worried about the pollution and waste arising from it is not a realistic solution under these circumstances.

While fewer and fewer people can go without a device like a cellphone, only a tiny number of people are directly involved in making decisions about how those devices are designed and manufactured in the first place. Consequently, there is a substantial difference in power between people who make design and manufacturing decisions about electronic devices such as cellphones compared to ordinary consumers of those devices so designed and made. As an everyday user of a device like a cellphone, one's power to make changes that would improve the pollution and waste consequences of electronic devices is quite limited. For example, ordinary people have little in the way of choice when it comes to the chemical composition of materials of devices, their durability, or their repairability. One can go into an electronics retailer and be presented with a rich menu of different makes and models, each with myriads of different functions. In that sense, choice is abundant. However, consumers have no meaningful choice over the pollution and waste implications of those same devices they might purchase. As is discussed in Chapter 2, by far the most waste and pollution arising from electronics occurs during the mining for their materials and during various manufacturing processes. These are stages in the life cycle of electronic devices over which consumers have little to no say. Despite the abundance of choices in terms of makes, models, and features, the underlying pollution and waste arising from all those different makes and models are so similar as to make the idea of consumer choice meaningless. Simply changing purchasing habits is insufficient on its own to overcome the problems of pollution and waste arising from electronics. Even if consumers repair or recycle their devices, that activity cannot recoup the magnitude of pollution and waste that arose before those devices were even purchased. As is discussed in Chapter 2, this situation does not mean recycling and consumer repair should be abandoned. They are, however, insufficient on their own as solutions to the overall problems of pollution and waste arising from electronics.

Strategies for Thinking Critically about E-waste

Statistics about waste are notoriously slippery. For example, estimates of total waste arising globally range between 2 billion and 32 billion metric tons annually (MacBride 2022). The upper bound of 32 billion is 16 times larger than the lower bound estimate of 2 billion. Clearly, there is a very wide range of uncertainty about the total amount of waste arising on the planet. One of the reasons such a wide range occurs is that municipal solid waste (MSW) is often relatively well measured in comparison to how waste arising from primary industry and manufacturing is measured. Data on MSW typically originate from actual weight measurements done as part of ordinary waste management activities such as collection and disposal. In contrast, waste arising from primary industry, such as mining and manufacturing, is often measured using estimates or models and, if the data are publicly reported at all, those reports tend to come from industries themselves rather than independent regulatory bodies (there are some exceptions to this rule discussed in Chapter 2).

Perhaps as a consequence of the higher quality documentation of MSW data, conversations about waste often elide MSW with total waste arising (MacBride 2022). That is, a considerable amount of waste arising at industrial sites is not subject to direct public measurement in part because there is often no legal mandate to do so. Even where such data collection is mandated, the data

Background and History

gathered and reported may be based on estimates or models rather than direct measurement. This situation can lead to very significant differences between the values of pollution and waste reported by the mining and manufacturing sectors when they are compared to direct measurements of waste, such as MSW. For example, a recent study that directly measured emissions from tar sands mining in Canada and compared them to values reported by industry found those direct measurements to "exceed oil sands industry-reported values by 1900% to over 6300%" (He et al. 2024, 1). The oil sands industry is an important source of fossil carbon-based raw materials used to make plastics, including those that are incorporated into many electronic devices. Despite this direct connection to "upstream" emissions like those reported by He and colleagues (2024), concerns about e-waste are usually focused on the "downstream" aspects of pollution and waste arising from electronics when consumers discard their devices. Yet, to elide post-consumer discards with pollution and waste attributable to electronics (i.e., "e-waste") means that all the "upstream" pollution and waste arising from mining and manufacturing for electronics are left out of the picture. The implications of widening the view of what counts as e-waste are discussed more fully in Chapter 2.

The slipperiness of waste statistics and the wide range of uncertainty they imply point to a need for strategies for thinking critically about knowledge claims about waste generally and e-waste specifically. Scholars in a field sometimes called Science and Technology Studies (STS) offer some approaches and concepts for doing so. A prominent theme in STS research is how knowledge in the sciences or in technical engineering comes to be settled as factual. The Latin etymology of the word "fact" points to an important claim made in some of this STS research. The word "fact" is based on the Latin root *facere* and, as such, it shares an etymological root with "artifact." In everyday conversation, "artifact" usually means something that is made by humans through some combination of creativity, ingenuity, and tools or instruments. A "fact," on the other hand, is supposedly free of all such human intervention. STS researchers, however, show repeatedly that the actual activity of doing science and engineering work that leads to factual knowledge requires all kinds of human creativity and ingenuity in combination with all sorts of technical instruments to build up and substantiate any claims that eventually come to be taken as facts. Consequently, rather than there being and incommensurable distinction between facts and artifacts, they are indeed quite similar. Artifacts are built and so are facts.

If this connection between artifacts and facts seems tricky, then consider the following analogy: a house is something that is built. Human labor is involved, usually by many different people with different specialized skills, as well as a range of materials such as wood, concrete, steel, and tools—everything from hammers and nails, to levels, or any of the other myriad implements needed to build a house. Technical knowledge and know-how are also important. The labor of many different people involved in building a house extends beyond the construction site itself to, for example, the manufacturing of the steel, the making of the lumber, not to mention all the labor involved in making the various fixtures and finishes. In this straightforward sense building a house is a collective effort. In other words, in a very mundane way, building a house is a social activity; it involves a collection of people and things to achieve its construction. In this way, it would be accurate to describe a house as something that is socially constructed. Sometimes the phrase "socially constructed" is used as a kind of accusation made by an accuser to criticize something for supposedly being "made up" as in the sense of being fake. Yet, it is not hard to understand that if someone pointed to an actual house and described it as being socially constructed that they are just offering an accurate description of the house and do not mean the house is fake. In this sense, STS researchers argue scientific and technical facts are also accurately described as being socially constructed, that is, they are constructed in the sense that houses are built.

This nuanced approach to understanding how claims to knowledge are built and come to be settled as factual offers an important strategy for thinking critically about knowledge claims about e-waste. It recognizes that all claims to knowledge are built or constructed. There is just no way to avoid relying on some collection of human creativity, know-how, and implements, instruments, or tools in attempts to make trustworthy knowledge about some phenomenon or another, including about e-waste. This kind of orientation to knowledge claims helps set up assessments of them in a different way than simply a binary between true or false, real or fake. A first step is to recognize that all knowledge is constructed. That means both claims and counterclaims to factualness are built or constructed, so constructedness is not a characteristic that can be used to distinguish one claim as true and another particular claim as false. Instead, a different staging of assessment is necessary: How well built is this or that claim to knowledge? Who builds it? Using which materials and instruments? Where? When? Under what conditions? Seeking out answers to these kinds of questions is a strategy to systematically assess which claims are built in more (or

Background and History

less) trustworthy ways. Claims to knowledge about e-waste—such as statistics about how much is produced, by whom, where, and under what conditions—are sometimes controversial. To weather a storm of controversy, it is important to assess the strength of the socially constructed facts one might take shelter in. Even a house built with a weak foundation or frame might provide some shelter from a light rain shower, but to weather a hurricane a house must be much better constructed.

E-waste: A Global Synopsis

The e-waste problem is global in extent, but the geography of the problem is uneven in its distribution of causes and consequences. It is also a problem that brings with it many challenges for characterizing its extent and impacts both quantitatively and qualitatively. One of those challenges is defining what counts as e-waste. It may be surprising to learn that something as basic as a definition of what counts as e-waste is difficult to find. However, when it comes to e-waste the "e" for "electronics" is particularly tricky. This is because "electronics" covers a very wide range of devices in different parts of people's daily lives. Some devices, like phones, laptops, and tablets are easy to understand as devices that could fall under a common-sense definition of an electronic device. But what about a washing machine or a refrigerator with wifi capabilities? How about an energy-efficient light bulb with a circuit board in it? Or a car, many contemporary versions of which, have more microchips in them than a typical desktop computer? In some parts of the world some of these devices are, in fact, included in laws that regulate the collection and disposal of e-waste. Yet, these same devices (and others) do not count in other places that also have e-waste legislation, such as some US states. Defining e-waste turns out to be surprisingly tricky. Definitions are just one of the challenges related to approaching e-waste as a pollution problem. A deeper discussion of those challenges is taken up later in the book. For now, it is possible to use data from a variety of sources to map some broad outlines of the e-waste problem.

Estimates of Total Mass and Growth Rates of E-waste Arising

Three aspects of the amount of e-waste arising globally are important to understand: the total quantity arising, the geographic variation of that total as it relates to different countries or regions around the world, and the degree to which e-waste is moved from its points of collection to where it is eventually processed for reuse and/or disposal. One authoritative source for statistics about total e-waste arising globally comes from the Global E-waste Monitor (GEM) (Baldé et al. 2024). GEM is a collaboration between several UN agencies, the International Solid Waste Association (a trade association), and the German Ministry for Economic Cooperation and Development.

According to the latest GEM report, 62 megatons (Mt) of e-waste were generated globally in 2022 (Baldé et al. 2024), up from 53.6 Mt in 2019. The increase in mass of e-waste arising between 2019 and 2022 represents an increase of over 15 percent over three years (Forti et al. 2020) or an average growth rate of about 5 percent per year in e-waste generated globally. A growth rate of 5 percent per year might sound small, but it means that if this growth rate remains constant then the total amount of e-waste arising globally will double its 2019 total to over 100 Mt sometime between 2030 and 2040. Table 1.1 provides data for total metric tons of post-consumer e-waste arising globally between 2014 and 2022.

While the total amount of e-waste arising globally is one part of the story, another key aspect to understand is geographic differences in the mass of e-waste arising. There are important differences between countries and regions of different income brackets as well as variations relating to relative economic wealth of people or households within different countries and regions. GEM data show that there are significant differences in e-waste arising between average households across countries in different income categories (high income, high-middle income, middle income, middle-low income, and low income). For example, in high-income countries, where average household size is 2.8 people, ownership of laptops averages 1.6 laptops per person. Meanwhile, in low-income countries where average household size includes five people,

Table 1.1 Total Post-Consumer E-waste Arising Globally, 2014–2022

Year	Total Post-Consumer E-waste (metric tons)
2014	41,800,000
2016	44,700,000
2019	53,600,000
2022	62,000,000

Sources: Data from Baldé et al. (2015); Baldé et al. (2017); Baldé et al. (2024); and Forti et al. (2020).

ownership of laptops averages only 0.1 per person. By contrasting the average households of the richest countries with the average households of the poorest countries in this example, it is quickly apparent that e-waste may be a global problem, but responsibilities for its creation are far from equally shared. Average households in high-income countries typically own more than ten times the number of laptops per person as does an average household in a low-income country. Interestingly, the differences in ownership between these two categories of households are less stark when it comes to cellphones: average households in high-income countries have 1.4 phones per person whereas average households in low-income countries have 0.6. These latter figures mean that average households in high-income countries "only" have about double the number of cellphones per person as do average households in low-income countries. These smaller differences between the richest and the poorest when it comes to phones speak to how important this category of device has become to daily life, albeit unevenly, in different countries around the world. Figure 1.1 provides an overview of post-consumer e-waste arising per person in different regions around the world. Data in the figure are for two periods, 2010 and 2022, which

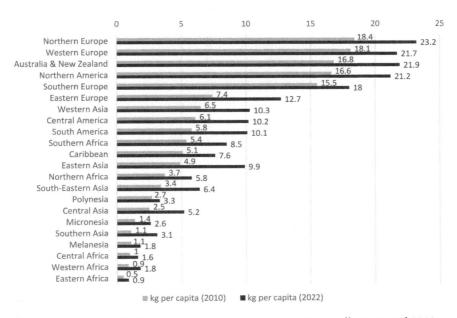

Figure 1.1 Per-Capita Post-Consumer E-waste Arising Regionally, 2010 and 2022
Source: Baldé et al. (2024).

provides some indication of change over time. The data also make it clear that different regions have different responsibilities with respect to contributions to overall post-consumer e-waste arising globally.

Beyond the broad income category differences just described, examining geographic differences can offer further insight into the uneven contributions of differently situated people and places to the overall global e-waste problem. Africa—with more than 1.1 billion people and fifty-four countries—generated 2,905 kilotons (kt) of e-waste in 2019 (or about 2.5 kg per capita) according to data from the Global E-waste Statistics Partnership (GESP) ("Global E-Waste Statistics Partnership" 2024). Contrast those figures with the Americas with less than 1 billion people in thirty-five countries and a total of 13,120 kt of e-waste (or 13.3 kg per capita)—more than four times the post-consumer e-waste generated per person in Africa—according to GESP data. Meanwhile, Asia's 4.4 billion people in forty-nine countries generated 24,896 kt, or 5.6 kg per capita, of e-waste in 2019—that is more total e-waste than in the Americas, but less than 50 percent of the Americas on a per capita basis. Europe's 740 million people in fifty countries generated 12,013 kt of e-waste or 16.2 kg per capita in 2019. Like the Americas—where the United States plays an outsized role—Europe generates four times more e-waste in total than Africa and almost 6.5 times more than Africa on a per capita basis; and Oceania with its 41.5 million people in fourteen countries generated a total of 667 kt of e-waste or 16.1 kg per capita in 2019.

These data convey several key issues about e-waste as a global pollution problem. First, the magnitude of the problem is measured in the millions of tons, or megatons, of waste arising. Second, the rate at which e-waste is being generated is increasing year-over-year at somewhere between 3 and 5 percent per annum. So, the e-waste problem is not just large in magnitude; its magnitude is also growing every year. Third, while the e-waste problem occurs around the world—that is, it can be found in any country—there are wide variations in the amount of e-waste arising when different groups of people and places are compared. This kind of variation points to a need to at least consider differential responsibilities for e-waste generation and how e-waste might be reduced or eliminated. Such variation also points to some of the problems built into comparative measures like per capita statistics whether they be about e-waste specifically or waste more generally.

Quantitative measures like per capita are often understood to be faithful representations of things that are not numbers, for example, a pile of discarded electronics weighs a certain number of kilograms. Yet things are not so simple. If that pile of electronics weighs a total of 5 kg of which 100 g are copper wires and another 100 g are mercury switches, these three measures are commensurable in terms of mass, but incommensurable when measured in terms of toxicity. Copper is fairly benign and presents a low risk of toxicity to people, except those who have rare genetic characteristics. Mercury, however, can be acutely toxic. The electronics discarded at a hypothetical curbside might have an equal quantity of copper and mercury, but the mass of those two metals means quite different things from a toxicological point of view. In this practical sense, numbers are not neutral or objective—which does not mean they are "bad"—but they do perform work. The work they perform relates to shaping a problem like e-waste in particular ways and not others. That matters because how a problem is shaped, or framed, makes certain ways of thinking about what to do about the problem possible, while taking other possibilities off the table all together. Per capita measures of waste perform work in a similar way (Liboiron 2013). For example, per capita measures of e-waste create an equivalence between individual people that poorly represents the wide variation in e-waste generation that exists between different groups of people and places when measured in terms of income. One of the risks of making such an equivalence is that "everyone" can seem to have the same responsibility in per capita terms. Yet, it would be absurd to suggest that a bank teller using a computer purchased by her employer and required for her work is equally as responsible for the waste arising from that computer when it is later disposed of by the bank that employs her. One of the actors in this latter scenario is an individual person, while the other is a corporate entity comprised of many people with massively more purchasing power than any one of its individual employees. These complexities around measuring waste in general and e-waste specifically on a per capita basis mean that it is important to keep in mind differential power relations whenever a waste problem is being analyzed, especially when solutions are being proposed. Complexities like this are addressed in more detail in Chapter 2.

Transboundary Shipments: Patterns and Processes

The sheer magnitude of e-waste arising is one facet of the global e-waste problem. Another important issue is the shipment of post-consumer e-waste across borders for reuse, processing, and disposal. Indeed, the issue of transboundary shipments of e-waste is repeatedly highlighted by concerned citizens, environmental non-government organizations (ENGOs), and media. GEM estimates suggest a range of 7–20 percent of total e-waste arising globally is shipped across borders (Forti et al. 2020, 55). That means that of the 53.6 Mt of e-waste estimated to arise globally in 2019, somewhere between 3.7 and 11.3 Mt were subject to transboundary movement. A more recent GEM study devoted specifically to transboundary shipments suggests 5.1 Mt, or about 10 percent of total e-waste arising globally, moved across borders in 2019 (Baldé et al. 2022, 8). The latest GEM report also puts the figure of transboundary movements of e-waste at 5.1 Mt, and further notes that approximately 3.3 Mt, or 65 percent of total transboundary shipments, are moved "through uncontrolled and undocumented" trade (Baldé et al. 2024, 15). GEM researchers are careful to note that their estimates probably under count the true figure for transboundary shipments of e-waste because of the lack of data and the challenge of universally accepted definitions of what counts as e-waste. Notwithstanding these inherent data challenges, GEM's figures are incorporated into policy negotiations under the Basel Convention—an international agreement that seeks to regulate and control transboundary shipments of hazardous wastes, including e-waste.

The significance of the Basel Convention is discussed in more detail later in this chapter. For now, it is important to recognize the importance of the percentage range (7–20 percent) of total e-waste arising that moves across borders according to GEM's analysis. If that range is correct, then somewhere between 80 and 93 percent of total e-waste arising does not move across borders. In other words, although transboundary shipments of e-waste are a matter of concern and receive a great deal of attention from engaged citizens, ENGOs, and media, this part of the e-waste problem is a relatively small proportion of the overall issue. However, it does not follow that just because a part of a problem may be smaller than other parts of it, that those smaller parts can or should be dismissed as unimportant. Also, framing the problem of e-waste strictly in terms of weight is insufficient if the problem as a whole is to be dealt with appropriately. Why? Electronics are composed of a multitude of materials. As mentioned in the previous section, some of those materials, like the copper in wires, are more or less benign from a toxicological point of view. But other materials are not. Some materials, like mercury that may be part of the soldering that attaches components to computer motherboards, can be acutely toxic. What that means is while 100 g of copper and 100 g of mercury are equivalent in terms of weight, they are totally different in their capacity to inflict toxic harm. So, what on the surface might seem like a simple problem—measuring e-waste—actually comes

Background and History 17

with difficult decisions and value judgments that play an important role in shaping how the problem of e-waste is understood. In turn, how a problem is understood plays a role in determining what solutions to that problem so framed might seem appropriate while also making alternative solutions more difficult or even impossible to imagine. This measurement challenge and how it relates to framing e-waste as a problem is taken up in detail in Chapter 2.

Some of the most common representations of the e-waste problem begin with images depicting trashed electronics, often in flames and tended to by people scavenging through the remains of digital devices. Some of the most common images purport to show people and places in Africa. Such images do show actually existing situations on the ground but do so in ways that erase underlying complexities that give rise to the situations depicted in these images. For example, although one can certainly find scrap heaps of discarded electronics in African locations, Africa exports more e-waste than it imports. Data from *resource.earth*, a project of an independent, UK-based policy institute called Chatham House, shows that Africa imported almost 324 million kg of e-waste between 2000 and 2020, while the continent exported almost 2.6 billion kg of e-waste over the same time period. At first this may seem surprising given that so much of the media attention on the global e-waste problem focuses on the issue of flows of e-waste from richer countries of the Global North to poorer countries of the Global South. Just because Africa exports more e-waste than it imports, that does not mean there is no need to properly address the hundreds of millions of kilograms of e-waste that does arrive in African destinations. But the situation does suggest that properly addressing the situation means approaching the problem differently than simply assuming it is one defined by a uni-directional flow of waste to Africa. These points are picked up in more detail in Chapter 2.

Researchers at Chatham House use data available from the United Nations COMTRADE database and perform a variety of statistical tests to ensure the data they make publicly available through the *resource.earth* project is of high quality (Chatham House, n.d.). Even with these quality checks, there are limitations to these data. For example, the data only cover scrap batteries and electric accumulators, excluding things like monitors, keyboards, and other peripherals. The data also exclude heavier items, such as washing machines, refrigerators, and AC equipment, that other sources of e-waste data, such as the Global E-waste Monitor, do count. So, the Chatham House data provide a partial picture—as all data do—of the transboundary shipments of e-waste.

From this partial picture the following broad patterns may be gleaned. Europe and North America are the regions with the largest transboundary flows of e-waste. Yet, there are important differences between these two regions. First, over 97 percent of Europe's transboundary flows of e-waste—over 7.5 billion kg imported between 2000 and 2020—circulate across borders within the European region. Most of these transboundary flows within the European region arrive in countries like Germany and Belgium. These countries are home to large industrial-scale recycling and secondary mineral processors that derive their feedstock from e-waste in the form of electronic scrap. Companies such as Arubis (Germany) and Umicore (Belgium) process electronic waste to recover metals such as aluminum, copper, and other precious and semi-precious metals. Also, counter to what might be expected, Europe's total imports of e-waste include over 68.6 million kg from countries of sub-Saharan Africa.

In contrast to the situation in Europe, most of North America's transboundary e-waste flows toward Latin America and the Caribbean. However, this inter-regional flow is dominated by exports from the United States to Mexico at over 4.5 billion kg (4.5 Mt) or more than 68 percent of all exports out of the North American region between 2000 and 2020. The next most important destination for exports from the North American region is South Korea. At over 1 billion kg (1 Mt), South Korea received almost 16 percent of total exports from the North American region between 2000 and 2020 and more than 94 percent of all those imports arrived from the United States. According to the US Geological Survey, both Mexico and South Korea are significant importers of metallic minerals for processing and rely on secondary minerals (i.e., minerals derived from recycling) for significant portions of that processing (United States Geological Survey 2022; 2023). E-waste, such as batteries, can be an important source of secondary minerals. For example, Mexico's lead refining industry processed an average of 253,000 metric tons of secondary lead annually between 2014 and 2018 compared to an average of almost 116,000 metric tons of primary lead over the same period. Copper is another important mineral found in some types of batteries and Mexico smelted and refined 5,000 metric tons of secondary copper annually over the same period. Processing of primary copper in Mexico is much more important than processing secondary copper, but 5,000 metric tons of secondary copper is still significant. Secondary mineral processing is also important in South Korea. For example, secondary copper accounted for about 20 percent of the country's total copper processing (smelting and refining) in any given year between 2015 and 2019. Meanwhile, secondary lead processing in

Background and History

South Korea accounted for an average of 380,000 metric tons annually between 2015 and 2019 versus an average of about 394,000 metric tons of primary lead over the same period.

Wastes Are Not, They Are Made

Erich W. Zimmerman was an economist who devoted his professional life to thinking about resources. Zimmerman argued that there is no such thing out there in the world that is inherently a "resource." His innovation was to shift thinking about resources away from their properties as objects to the functional role(s) they play as means of satisfying human desires. And, since human desires can shift from time to time and place to place, what is considered a resource in one time and place may not be so in a different place or at another time. An aphorism for this idea attributed to Zimmerman goes as follows: resources are not, they are made. While Zimmerman was publishing his work in the early to mid-twentieth century, contemporary debates about shifting current industrial activities toward a circular economy echo some of his ideas. Advocates of circular economies claim that wastes can be transformed into resources. Historical shifts in how discarded electronics are portrayed bear out Zimmerman's ideas in some respects.

In the 1970s, the US Bureau of Mines published a series of studies investigating discarded electronics as a potential source of precious and semi-precious metals. These reports offer technical analyses of how metals could be recovered from these discards. What is striking is that the word "waste" never appears in relation to the electronics being discussed in these reports. Instead, the discarded electronics being studied are referred to as "scrap" (Dannenberg, Maurice, and Potter 1972). Indeed, the premise of these reports from the Bureau of Mines is that electronic scrap could be used—that is, it could function—as a resource for obtaining desirable metals like gold, silver, palladium, and other metals of interest. These reports offer a practical working out of Zimmerman's aphorism about resources. Resources come into existence in a functional way in relation to human desires to reach particular ends.

The Bureau of Mines reports understood discarded electronics as resources. It would be almost thirty years before concerns about discarded electronics would come to be framed as a waste management issue. One of the first mentions of discarded electronics as a problem for waste management came from a journalist writing in 2000 for a trade publication called *Electronic Business* (Arensman 2000). The journalist's article profiled recycling initiatives by electronics manufacturer Hewlett-Packard (HP) of, "assorted computers, printers, circuit boards and other technology industry refuse" (Arensman 2000, 108). Unlike the reports from the Bureau of Mines, the worries expressed in this business press article are about electronic waste taking up space in landfills and having potentially negative toxic consequences. What is important to recognize is that the very same collection of things—computers, printers, circuit boards, etc.— have completely different meanings in different times and different situations. In the 1970s, they were "electronic scrap," but by the 2000s they were "electronic waste." The shifting meanings attached to the same collection of things shows that those things do not have essential characteristics as either resource or waste. Instead, things deemed to be a resource or a waste are *made* to have those meanings and to be treated as such depending on changeable circumstances. Zimmerman's aphorism about resources holds for resources as well as waste: wastes are not, they are made.

Once "waste" is understood as a particular way of relating to people, places, or things labeled as such, it becomes apparent that waste problems are more than just technical problems in search of technical solutions. Indeed, scholars who investigate the social processes by which people, places, or things come to be understood and treated as waste argue that in addition to the technical questions of waste management it is also important to be mindful of the allegorical aspects of waste. Allegories are stories that can be interpreted to have moral and political meetings underneath their surface details. E-waste is a category of waste objects where allegorical storytelling is a core facet of the overall e-waste problem.

In 2002, a US-based ENGO published a report that focused on e-waste found in Asia that had originated in the United States. The name of the ENGO is the Basel Action Network (BAN), and it is named after the Basel Convention mentioned earlier. The report released by the group, called *Exporting Harm*, is a touchstone in the public understanding of the e-waste problem. *Exporting Harm* presents the e-waste problem as both a practical issue of discarded machines and an issue with moral and political connotations. For example, the cover of the report shows a young, brown-skinned child sitting on top of a pile of junk electronics; the child looks directly into the camera and, thus, directly into the eyes of a report reader. It is an image that is meant to provoke an emotional response. E-waste in such a context is no mere neutral pile of "stuff." Instead, it is stuff rife with emotional content and meaning as well as being a technical waste management problem. As such, the complexities of the e-waste problem multiply.

Between 2000 and 2002, e-waste had been made into a problem different from its earlier characterization as "electronic scrap" in the 1970s. Not only had discarded electronics become "e-waste," but they were also redefined as reflecting a particular kind of waste problem—one that arises after consumers get rid of their devices. The post-consumer framing of the e-waste problem has quite important implications which, though they may not be intentional, nevertheless strongly shape the waste management responses to the problem so framed. For example, one of the solutions to the e-waste problem when it is understood to be a post-consumer problem is to offer up post-consumer recycling as a solution. While there is nothing inherently wrong with post-consumer recycling, it cannot cope with the magnitude of pollution and waste arising from earlier stages necessary for making electronics, such as mining and manufacturing. This issue is taken up in detail in Chapter 2. For now, the important point to take away from this section is that the shifting understandings of discarded electronics—as scrap, waste, resource—demonstrate that the meaning these things take on is neither essential, nor fixed. Understanding this is important for enabling the imagining of a broader range of potential responses to the problem of waste arising from electronics.

E-waste Regulation around the World

Efforts to regulate post-consumer e-waste have been ongoing since 1992 when South Korea became the first country to promulgate laws covering the recycling of televisions and washing machines (Jang 2010). As of 2022, eighty-one countries around the world have passed legislation covering the management of e-waste (Baldé et al. 2024). Despite diversity in the details of the growing number of legislative attempts to manage e-waste, there is a key commonality shared by much of the e-waste legislation passed in different jurisdictions around the world: an almost exclusive focus on e-waste as an "end-of-pipe" or "post-consumer" waste management problem. In this framing, waste from electronics happens only after businesses, consumers, or households get rid of them. However, most of the pollution and waste arising from electronics happens during the mining and manufacturing phases of their existence. One important consequence of the post-consumer focus of most e-waste legislation is that all of the pollution and waste arising from mining and manufacturing associated with electronics are bracketed out of consideration. This bracketing out has important consequences in terms of what solutions to the problem so framed are thought to be relevant.

A second major commonality in e-waste regulations, although more recent in its emergence, is an emphasis on flow control. In waste management contexts, flow control refers to regulatory requirements that a given waste stream be delivered to specific facilities in a given jurisdiction. In the context of e-waste management, an additional layer of meaning has become more relevant as concerns over access to critical minerals grow. E-waste can be processed to extract metals and other materials crucial for technologies necessary to broad-scale shifts in energy infrastructure toward electrification. These same metals and materials often have strong geopolitical chokepoints that some countries and regions understand as threats to their security. Thus, flow control as it relates to e-waste is increasingly about controlling access to the critical materials embodied in discarded electronics, which can, at least in principle, be extracted for reuse in new industrial production. E-waste laws in the European Union (EU) and in member countries of the Organization for Economic Cooperation and Development (OECD) tend to attempt to hold discarded electronics within their borders for scrap recovery. In contrast, China—which exercises significant control over the mining and processing of some strategic metals—must import vast amounts of feedstock for its industrial manufacturing infrastructure that exports many products to the world including consumer electronics. Thus, in contrast to the EU and OECD approach to managing e-waste, China's Belt and Road initiative—a massive infrastructure development program—can be interpreted as infrastructure facilitating the flow of products discarded elsewhere as sources of scrap materials for feedstock into China's industrial production. The two different approaches to flow control are perhaps best illustrated by the EU's and the OECD's tendency to rely on the Basel Convention, which attempts to halt the transboundary flow of e-waste (and other hazardous wastes). Meanwhile, although China is a party to the Basel Convention, in attempting to control what it deems to be imports of wastes to its shores, it tends to rely on the World Trade Organization (WTO). These distinctions are examined in more detail in the following sections.

The Basel Convention

The Basel Convention on the Control of Transboundary Movements of Hazardous Wastes and Their Disposal (hereafter, the Convention) is an international treaty designed to regulate transboundary shipments of waste. E-waste is one category of waste that the Convention attempts to regulate. As of 2024, there are 191 countries or territories that are signatories to the Convention (the United States has signed, but not ratified the treaty). The Convention was adopted by delegates in 1989 in the Swiss city of Basel, and the agreement came into force in 1992.

E-waste became a category of special concern for the Convention in the years since its adoption. A technical committee of the Convention was convened in 2010 to make a distinction between waste and non-waste electronic equipment in an attempt to bring regulatory clarity to existing definitions in the treaty. Technical guidelines on the distinction between waste and non-waste electronics were adopted on an interim, non-legally binding basis in 2015 after five years of debate. Since then additional amendments to the Convention have been made in an attempt to better manage transboundary flows of e-waste, and those changes will come into effect in 2025 (Dehmey and Puckett 2024). It remains to be seen, however, whether these changes will resolve the challenges of distinguishing between waste and non-waste electronics under the Basel Convention.

It may seem strange that after more than a decade of negotiations technical experts have still not been able to come to a consensus on a distinction between waste and non-waste electronics. Yet, the ensuing intractability stems from fundamental challenges of defining "waste" and "hazardousness" in universally agreed-upon ways. For example, there can be regulations around chemicals known to have toxic harms to people and environments, such as polychlorinated biphenyls (PCBs). But such regulations can differ by orders of magnitude between jurisdictions. For example, regulations in the United States permit concentrations of PCBs that are five times higher than those permissible in the United Kingdom. Although the US allowable limits are much higher, the regulations in both jurisdictions permit a certain amount of harm and there is nothing in the physical characteristics of the chemicals involved that would provide neutral criteria by which to judge one of those regulations correct and the other wrong (Wynne 1987). Similar challenges are inherent in trying to draw a line between waste and non-waste electronics, in part because those devices are composed of many materials that can be toxic.

In addition to the challenges of defining "hazardousness," the Convention also permits export of discarded electronics for "direct reuse" (Basel Secretariat 2019, 61). Challenges of definition and interpretation arise here as well because the Convention states that reuse can include "repair, refurbishment or upgrading, but not major reassembly" yet it provides no definition of what counts as such

24   *Electronic Waste*

major reassembly. Thus, what remains at issue under the Basel Convention is how to distinguish legitimate from illegitimate transboundary shipments of discarded electronics.

The Bamako Convention

The Bamako Convention is an international treaty among twenty-five African countries that, like the Basel Convention, regulates transboundary movement of hazardous wastes. Negotiated in Bamako, Mali, the impetus for the Bamako Convention arose directly from some participants' frustrations over the shaping of provisions of what would become the Basel Convention (Clapp 1994). Indeed, Bamako was signed just two years after the Basel Convention and Bamako came into force in 1998, just a few years after the Basel Convention did. Unlike Basel, Bamako explicitly attempts to create a regulatory regime organized around the "precautionary principle" and "clean production" practices (Bamako Secretariat 1991). In this case, the precautionary principle is an orientation toward regulation that significantly limits or outright prohibits the use of chemicals and other materials in manufacturing that are likely to cause toxic harm, even if definitive proof of such harm has not yet been established. "Clean production" in the context of the Bamako Convention is a requirement for pollution control strategies to avoid a sole focus on "end-of-pipe pollution control such as filters and scrubbers, or chemical, physical or biological treatment" (Bamako Secretariat 1991). In these respects, the Bamako Convention is very different in its regulatory orientation than that of the Basel Convention. Bamako focuses on changing how products that will eventually be discarded by consumers are made in the first place. In contrast, the Basel Convention focuses on what to do with wastes after they already exist.

Although there is no explicit reference to electronics in the Bamako Convention, it does prohibit the transboundary trade of wastes containing a variety of metals and metal bearing compounds relevant to electronics. Even so, the provisions of Bamako regarding the precautionary principle and clean production make this convention worthy of consideration for lessons it offers relevant to the elimination and mitigation of the e-waste problem. The contrast of Bamako with Basel shows, once again, that the meaning and practical realities of what waste is and how it is constituted are characteristics that are made, not essential characteristics inherently part of manufactured products that, following use, automatically fall into the category of waste. Bamako makes waste differently than Basel by focusing regulatory attention "upstream" on manufacturing, that is, the beginning of the pipe rather than its end. Discussions later in the book on the relative contribution of mining and manufacturing to total pollution and waste arising from electronics offer a reminder of why Bamako's approach may be more relevant for articulating solutions to overall pollution and waste problems arising from electronics than the Basel Convention's end-of-pipe regulatory approach.

Africa and Middle East/West Asia

As many as twelve countries in Africa have legislation specific to e-waste but in most cases these laws are currently in draft form as of 2023 (Forti et al. 2020; "Global E-Waste Statistics Partnership" 2024). Ghana and Nigeria are two exceptions, having passed e-waste legislation in 2016 and 2011, respectively. Under Ghanian law e-waste is defined via a voltage route like the path taken in the EU (see section on Europe). Nigeria forgoes voltage but takes a categorical approach to defining e-waste that is also like that of the EU (i.e., large household appliances, small household appliances, IT and telecommunications equipment, etc.). Both countries' e-waste legislation attempts to implement a form of polluter-pays-principle. The intent of such a principle is that manufacturers or importers of electronic equipment hold responsibility for paying for the cleanup costs of that equipment. This principle is a common feature of e-waste legislation in many jurisdictions, not just Ghana and Nigeria. A deeper discussion of what is called extended producer responsibility (EPR) and the polluter-pays-principle can be found in Chapter 2.

The only country in the Middle East/West Asia with e-waste legislation currently in force is Israel. The country's law came into effect in 2014 and covers batteries, mobile phones, computers, televisions, and refrigerators (Government of Israel 2020). The legislation was passed in part as a requirement of Israel joining the Organization for Economic Cooperation and Development (OECD) (Davis and Garb 2019a). Israel's legislation, like that of many other jurisdictions, is premised on the idea of EPR (see Chapter 2 for further discussion). Moreover, as in other jurisdictions in which "the informal sector" is an integral part of e-waste collection and processing, the Israeli system depends on deeply unequal power relations between the Israeli state and its citizens and Palestinians in the Occupied Territories, where substantial processing of e-waste sourced in Israel occurs (Davis and Garb 2019b).

Americas

The North American Free Trade Agreement (NAFTA, now the Canada-United States-Mexico Agreement or CUSMA) covers Canada, Mexico, and the United States. Although these countries form a trade bloc, each of the three countries approaches e-waste somewhat differently. None of the NAFTA/CUSMA countries have federal legislation specifically about e-waste. Canada and the United States rely on sub-national jurisdictions to provide that legislation. Mexico, meanwhile, has country-wide legislation covering "special wastes." The "special wastes" category does cover some categories of electronics but only for generators of at least 10 tons annually, which would typically exclude private citizens and households.

E-waste legislation in Canadian provinces and territories and US states creates a patchwork system of laws covering discarded electronics in the two countries. In the Canadian case, the early intervention of a nonprofit trade association called Electronics Product Stewardship Canada (EPSC) comprised of several major electronics brands helped to create a more consistent legislative landscape regulating e-waste than that of the United States. EPSC began its work in 2004 and soon joined forces with retailers to form a nonprofit organization called the Electronic Products Recycling Association (EPRA). EPRA oversees the e-waste recycling systems of all provinces, except for Alberta's public authority and those of Northwest Territories and Yukon (Electronics Recycling Association 2014). EPSC/EPRA advocated for a product category approach to defining e-waste and wrote model legislation that was later adopted in all provinces (except for Alberta, Northwest Territories, and Yukon). The EPSC/EPRA model e-waste regulation covers equipment such as display devices, telephones, home and personal A/V equipment, computers, computer peripherals (e.g., keyboards, printers, and mice), and even electronic musical instruments, and medical monitoring equipment.

The Canadian system is financed through the collection of fees at the point of purchase by consumers. These fees go by different names. They are sometimes called "advance disposal fees," "environmental handling fees," or "eco fees," but they all work in the same general way. These fee systems collect a pool of money which is then used to pay for the cost of collecting and recycling electronics later discarded by consumers. Paying for collection and recycling using such fees is a common approach in many jurisdictions and is often portrayed as a form of EPR. However, such a description is arguably a misnomer since it is consumers, rather than producers, who actually pay these fees. The deeper implications of

Background and History 27

financing EPR systems in this manner are discussed in more detail in the next chapter.

E-waste legislation in the US case is the most fragmented amongst NAFTA/CUSMA countries. At the federal level is an Obama-era policy called the National Strategy for Electronics Stewardship (NSES) (United States Environmental Protection Agency 2011). NSES is more aspirational policy than legally enforceable legislation. It recommends incentivizing "green" design choices around things such as energy efficiency and the upgrade ability of devices. NSES also recommends the federal government show leadership by, for example, prioritizing surplus federal electronics for reuse in schools and nonprofits; and ensuring responsible recycling by using certified recycling contractors and avoiding the export of non-working electronics.

In 2003, California became the first state to introduce e-waste legislation. Over the next decade, twenty-five more states would follow, with the most recent legislation passed in 2014 for the District of Columbia (Electronics Recycling Coordination Clearinghouse 2021). Two key characteristics of the US states' approach to e-waste regulation are important to highlight. First is the lack of consistency in what products are covered by state legislation. A continuously updated state-by-state list of products covered by e-waste laws is maintained by the Electronics Recycling Coordination Clearinghouse, an initiative of the National Center for Electronics Recycling (NCER) and Northeast Recycling Council (NERC). Although states amend and revise their e-waste legislation over time to include some previously excluded categories and vice versa, the overall picture of heterogeneous definitions of what counts as e-waste in different states remains the norm (Electronics Recycling Coordination Clearinghouse 2024). Second, the financing of the state systems is largely borne by consumers despite explicit regulatory references to EPR. The net effect of how financing is organized in these states' systems for collecting and recycling e-waste means that the costs of end-of-life management are largely kept externalized from manufacturers. Such cost externalization nullifies the price signal that is supposed to incentivize manufacturers to redesign their devices to reduce costs associated with waste.

Asia Pacific

Australia

Australia's approach to e-waste has been evolving over the last decade. Its early efforts focused on categorical definitions of e-waste covering batteries, air conditioners, and refrigerators (Australian Government, Department of Agriculture, Water and the Environment 2013). In 2016–17, the list of covered equipment expanded to televisions, printers, and computers under the National Television and Computer Recycling Scheme (NTRS).

In 2020, Australia passed the Recycling and Waste Reduction Act (RWRA) (Australian Government 2020). An organization called the Australia New Zealand Recycling Partnership (ANZRP) emerged out of this legislative revamp. ANZRP is an industry-led nonprofit managed by an eight-member board (ANZRP 2019). Five of the eight board members hale from large electronics manufacturers (HP, Dell, Fuji, Toshiba, and Canon). Dias, Bernardes, and Huda (2018) claim that industry finances the Australian e-waste management system but note the system has been criticized for its lack of transparency around how the financing actually works. Such a criticism suggests that, as in other jurisdictions (e.g., Canada), industry has found ways to keep the costs of product stewardship externalized from its bottom line and nullify any price signal that might incentivize upstream design changes that could mitigate post-consumer e-waste.

China

China has figured prominently in e-waste controversies since the early 2000s when Guiyu, a city in Guandong province, became the subject of a controversial report on e-waste exports from the United States to China by an American ENGO. In July 2017, China took the notable step of banning the import of materials it deemed "foreign garbage." In so doing the ban defined four material categories and twenty-four specific kinds of materials, some of which overlap with elements of e-waste (e.g., plastics like those used to encase devices) (Ministry of Environmental Protection of the People's Republic of China 2017). Notably, the ban was enacted by China through the WTO, rather than the Basel Convention, which raises questions about which of the two international agreements China understands to be more effective for regulating international flows of waste.

Much of the reporting on the 2017 ban enacted by China paid attention to its knock-on consequences for municipal recycling systems in wealthier countries in Asia, Europe, and North America. Municipal recycling systems in countries such as the United States, Canada, the United Kingdom, and Japan seemed to suddenly lose the single largest buyer—China—of the materials collected in those and other countries for recycling. The situation had the effect of significantly reshaping the global market for recyclable materials, especially plastics, including those derived from discarded electronics (Reed, Hook, and Blood 2018). Less attention in the media around this ban was paid to how it affected manufacturers in China. According to Bloomberg journalist Adam Minter, manufacturing interests in China advocated for a loosening of the restrictions under the "foreign garbage" ban because their industries are heavily dependent on feedstock materials derived from overseas recycling markets (Minter 2020). For example, at least 15 percent of China's annual aluminum production is derived from secondary scrap; meanwhile, scrap copper accounts for 20 percent of China's copper smelting and ranges from 40 to 60 percent of its copper refining (Xun 2018). Discarded electronics are an important source of these and other secondary scrap materials, including plastics.

China's e-waste regulatory framework focuses on domestic collection and eliminating imports (e.g., from the EU). China focused first on reducing and eliminating transboundary flows of e-waste into its territory. These efforts began in the 1990s and focused on product categories to define e-waste (e.g., TVs, refrigerators, air conditioners, microwave ovens, and computers) (Schulz and Steuer 2017). Not until 2009 did Chinese policymakers turn their attention to domestically generated e-waste. These newer regulations take a similar product category approach to defining e-waste as that of China's earlier legislative precedents for controlling imports of materials derived from discarded electronics.

Japan

Japan's e-waste legislation is framed within a broader circular economy policy framework that began to be articulated in 1999, well ahead of similar approaches in Europe and elsewhere (Japan, METI Ministry of Economy, Trade and Industry 2020). Japan's legislation on e-waste emerged in the late 1990s with the Act on Recycling of Specified Home Appliances (ARSHA). ARSHA takes a product category approach to defining what counts as e-waste. When the law came into effect in 2001 it covered a relatively narrow range of equipment— air conditioners, televisions, refrigerators and freezers, washing and drying machines.

Since the passage of ARSHA, Japan also amended its Resource Recycling Promotion Law of 1991. The amendment is called the Law for the Promotion of Effective Utilization of Resources (LPEUR) and passed in 2000 (Japan, METI Ministry of Economy, Trade and Industry 2007). LPEUR added personal computers and microwave ovens to the list of products covered. More recently, the Act on Promotion of Recycling of Small Waste Electrical and Electronic Equipment (PRSWEEE) came into effect in 2013 (Terazono 2013). PRSWEEE defines e-waste using twenty-eight categories of home appliances, several of which overlap ARSHA and LPEUR and adds products such as phones, cameras, vacuums, and rice cookers. A notable difference of the PRSWEEE law to that of other national-level e-waste legislation elsewhere is how it empowers municipal governments to define which specific categories of electronics those municipalities will collect within their own jurisdiction.

Singapore

Singapore was an early adopter of legislation to regulate toxic components of batteries—a key category under the broader umbrella of e-waste. The city-state passed legislation mandating the reduction of mercury content of batteries in 1992. Although confined to a single category of device—batteries—this legislation is important for its role in reducing pollution arising from electronics before they are discarded by their downstream users, that is, by consumers. Such an upstream approach can reduce the negative pollution consequences of later consumer behavior no matter what they do with those devices they toss away. Reducing the toxicants in batteries will not, of course, solve the problem of e-waste in a holistic sense. However, Singapore's approach to regulating e-waste upstream of consumers points to an important idea for e-waste management specifically and waste management more generally: being explicit about what theory of change a given management tactic is premised on. As discussed in Chapter 2, downstream approaches to waste management, such as post-consumer recycling, cannot recoup the upstream pollution and waste arising during the mining for, and manufacturing of, consumer products such as electronics.

Singapore began to regulate a range of post-consumer e-waste categories more recently under two policies: the National Voluntary Partnership for E-waste Recycling (NVPER) and the EPR System for E-waste Management (Singapore National Environment Agency 2019, 2021). These policies define e-waste via device categories. They include various types of electronics such as printers, computers, tablets, monitors as well as certain large appliances such as refrigerators, air conditioners, washing machines, but also e-bikes and scooters. NVPER is a voluntary program as the title of the policy indicates, but it is intended to become mandatory in 2024. Meanwhile, Singapore's EPR system covers consumer and non-consumer e-waste as separate waste streams. Singapore's EPR system is financed through subsidies to businesses that become certified under voluntary standards for the collection and processing of e-waste. As discussed in Chapter 2, this approach to EPR has the drawback of potentially neutralizing any price signal felt by manufacturers of electronics, thereby failing to institute genuine producer responsibility.

South Korea

South Korea was among the earliest countries to have legislation for e-waste. Beginning in 1992 the country had legislation on the books to cover televisions and washing machines. A decade later new laws expended the range of equipment covered and introduced an EPR regime.

By 2007, South Korea was already covering both electronic equipment and vehicles, something still rare across other jurisdictions. The country's Act for Resource Recycling of Electrical and Electronic Equipment and Vehicles (RREEV) defines e-waste as any "equipment or device (including components and parts thereof) operated by electric currents or electromagnetic fields" (Yun and Park 2007, 5). As contemporary commentators note, cars have increasingly become "rolling computers" (Patel 2017). South Korea's regulatory regime has been leading the way in this respect. However, as in other jurisdictions, the EPR aspects of the Korean legislation at least partially negate the price signal that genuine EPR would send to electronics manufacturers to incentivize the mitigation or elimination of pollution and waste arising from their activities.

Europe

EU-wide regulation of e-waste—called the Waste Electrical and Electronic (WEEE) Directive—came into effect in 2003. The EU takes an approach to defining e-waste that combines voltage-based criteria with size and categorical characteristics (e.g., large and small household appliances, IT equipment, tools, toys, and medical devices, among others). Due to the voltage-based approach to defining e-waste, EU laws cover a wider range of equipment than is typical in other jurisdictions. However, a distinguishing feature of the EU legislation is that it specifically excludes vehicles (except e-scooters).

Two other pieces of EU legislation influence the toxic qualities of post-consumer e-waste arising in the region. These are the Reduction of Hazardous

Substances (RoHS) and Registration, Evaluation, Authorisation and Restriction of Chemicals (REACH) regulations. Neither RoHS nor REACH is intended to regulate the collection and processing of discarded electronics. Instead, each of these laws regulates the use of chemical toxicants in the manufacturing stage and their presence in device hardware. RoHS focuses specifically on electronics and is aimed at reducing or eliminating the use of various heavy metals, such as lead, mercury, and cadmium as well as reducing the use of various flame retardant chemicals. REACH goes even further to cover all manufactured commodities in the EU. What is important about RoHS and REACH is how they intervene to change the very nature of waste before it arises after consumers get rid of devices. While this legislation will not reduce the mass of post-consumer e-waste, they do reduce or eliminate the toxicity of that e-waste so arising.

The EU legislation regulating WEEE management is financed through a version of EPR. Like other jurisdictions already described, the EU EPR system is structured in such a way that the costs of managing end-of-life electronics are externalized from the manufacturers of those products. Indeed, the WEEE regulations explicitly state their goal of shifting the "payment for the collection of this waste from general tax payers to the consumers of EEE, in line with the 'polluter pays' principle" (European Union 2012, OJ L:4). As such, this WEEE legislation formats consumers of electronics as the polluters, rather than manufacturers of those devices. Arguments can be made that a specific class of consumers should bear special responsibility for the end-of-life pollution arising from the products they consume. At the same time, such a framing means that actors with the most power to change overall pollution and waste arising— manufacturers—have their responsibilities minimized or eliminated.

Conclusion

Since emerging as a matter of concern in the early 2000s, global volumes of electronics discarded by consumers have grown significantly. With those concerns came attempts to manage and regulate this waste stream. The management systems that emerged continue to focus almost exclusively on e-waste as a post-consumer problem. With some variations in different jurisdictions around the world, most of the regulatory effort continues to focus on post-consumer recycling of discarded devices. Less attention is paid to interventions that might be made upstream during mining for, and manufacturing of, devices that would eventually become waste, but this is changing. The next chapter examines some of the problems and controversies related to e-waste when it is understood as a post-consumer problem in isolation from the pollution and waste consequences of the mining and manufacturing needed to bring electronic devices into being.

References

ANZRP. 2019. "Australian and New Zealand Recycling Platform Annual Report 2018–19." http://www.environment.gov.au/system/files/resources/8964a490-c173-44b5-b9d0-a0a1eebdec55/files/anzrp-annual-report-2018-19.pdf.

Arensman, Russ. 2000. "Ready for Recycling?" *Electronic Business*, November.

Australian Government. 2020. "Recycling and Waste Reduction (Consequential and Transitional Provisions) Act 2020." Attorney-General's Department. https://www.legislation.gov.au/Details/C2020A00120/Html/Text, http://www.legislation.gov.au/Details/C2020A00120.

Australian Government, Department of Agriculture, Water and the Environment. 2013. "Product List and Notices: 2013–14 Product List." Department of Agriculture, Water and the Environment, June 30, 2013. http://www.environment.gov.au/.

Baldé, C. P., F. Wang, R. Kuehr, and J. Huisman. 2015. "The Global E-Waste Monitor 2014: Quantities, Flows and Resources." Bonn: United Nations University.

Baldé, C. P., V. Forti, R. Kuehr, and P. Stegmann. 2017. "The Global E-Waste Monitor 2017: Quantities, Flows and Resources." Bonn: United Nations University. http://collections.unu.edu/eserv/UNU:6341/Global-E-waste_Monitor_2017__electronic_single_pages_.pdf.

Baldé, Cornelis Peter, E. D'Angelo, V. Luda, O. Deubzer, and R. Kuehr. 2022. "The Global Transboundary E-Waste Flows Monitor 2022." United Nations Institute for Training and Research (UNITAR), Bonn, Germany. https://ewastemonitor.info/wp-content/uploads/2022/06/Global-TBM_webversion_june_2_pages.pdf.

Baldé, Cornelis Peter, Ruediger Kuehr, Tales Yamamoto, Rosie McDonald, Elena D'Angelo, Shahana Althaf, Garam Bel, et al. 2024. "The Global E-Waste Monitor 2024." Geneva/Bonn: International Telecommunication Union (ITU) and United Nations Institute for Training and Research (UNITAR). https://ewastemonitor.info/wp-content/uploads/2024/03/GEM_2024_18-03_web_page_per_page_web.pdf.

Bamako Secretariat. 1991. "Bamako Convention on the Ban of the Import into Africa and the Control of Transboundary Movement and Management of Hazardous Wastes within Africa." http://www.au.int/en/content/bamako-convention-ban-import-africa-and-control-transboundary-movement-and-management-hazard.

Basel Secretariat. 2019. "Basel Convention: Revised 2019." http://www.basel.int/Portals/4/download.aspx?d=UNEP-CHW-IMPL-CONVTEXT.English.pdf.

Chatham House. n.d. "Resource.Earth." Accessed January 8, 2024. https://resourcetrade. earth/.

Clapp, Jennifer. 1994. "Africa, NGOs, and the International Toxic Waste Trade." *The Journal of Environment & Development* 3, no. 2: 17–46. https://doi. org/10.1177/107049659400300204.

Dannenberg, R. O., J. M. Maurice, and G. M. Potter. 1972. "Recovery of Precious Metal from Electronic Scrap." Report No. 7683. Salt Lake City, UT: United States Department of the Interior, Bureau of Mines.

Davis, John-Michael, and Yaakov Garb. 2019a. "Extended Responsibility or Continued Dis/Articulation? Critical Perspectives on Electronic Waste Policies from the Israeli-Palestinian Case." *Environment and Planning E: Nature and Space* 2, no. 2 (April 2019): 368–89. https://doi.org/10.1177/2514848619841275.

Davis, John-Michael, and Yaakov Garb. 2019b. "Participatory Shaping of Community Futures in E-Waste Processing Hubs: Complexity, Conflict and Stewarded Convergence in a Palestinian Context." *Development Policy Review* 37, no. 1: 67–89. https://doi.org/10.1111/dpr.12333.

Dehmey, Corey, and Jim Puckett. 2024. "In Our Opinion: Industry Must Commit to Basel e-Plastics Compliance." *E-Scrap News* (blog), February 1, 2024. https:// resource-recycling.com/e-scrap/2024/02/01/in-our-opinion-industry-must-commit-to-basel-e-plastics-compliance/.

Dias, Pablo, Andréa Moura Bernardes, and Nazmul Huda. 2018. "Waste Electrical and Electronic Equipment (WEEE) Management: An Analysis on the Australian E-Waste Recycling Scheme." *Journal of Cleaner Production* 197 (October): 750–64. https://doi. org/10.1016/j.jclepro.2018.06.161.

Electronics Recycling Association. 2014. "Who We Are." https://epra.ca/who-we-are.

Electronics Recycling Coordination Clearinghouse (ERCC). 2021. "Map of States with Legislation." https://www.ecycleclearinghouse.org/contentpage.aspx?pageid=10.

Electronics Recycling Coordination Clearinghouse (ERCC). 2024. "Maps." https://www. ecycleclearinghouse.org/maps.aspx.

European Union. 2012. *Directive 2012/19/EU of the European Parliament and of the Council of 4 July 2012 on Waste Electrical and Electronic Equipment (WEEE). Text with EEA Relevance. 197.* Vol. OJ L. http://data.europa.eu/eli/dir/2012/19/oj/eng.

Forti, Vanessa, Cornelis Peter Baldé, Ruediger Kuehr, and Garam Bel. 2020. "The Global E-Waste Monitor 2020: Quantities, Flows, and the Circular Economy Potential." Bonn/Geneva/Rotterdam: United Nations University. https://www.itu.int/en/ITU-D/ Environment/Documents/Toolbox/GEM_2020_def.pdf.

Gidwani, Vinay. 2012. "Waste/Value." In *The Wiley-Blackwell Companion to Economic Geography*, edited by Trevor J. Barnes, Jamie Peck, and Eric Sheppard, 275–88. Chichester: John Wiley & Sons.

Giridharadas, Anand. 2010. "Where a Cellphone Is Still Cutting Edge." *New York Times*, April 9, 2010, sec. Week in Review. http://www.nytimes.com/2010/04/11/ weekinreview/11giridharadas.html.

"Global E-Waste Statistics Partnership." 2024. *E-Waste* (blog). 2024. https://globalewaste.org/.

Goldstein, Jesse. 2013. "Terra Economica: Waste and the Production of Enclosed Nature." *Antipode* 45, no. 2: 357–75. https://doi.org/10.1111/j.1467-8330.2012.01003.x.

Government of Israel. 2020. "Bottles, Tires, Packaging, Appliances, Plastic Bags: Extended Producer Responsibility." GOV.IL, October 20, 2020. https://www.gov.il/en/Departments/Guides/extended_producer_responsibility?chapterIndex=4.

Gregson, Nicky. 2023. *The Waste of the World: Consumption, Economies and the Making of the Global Waste Problem*. Bristol: Bristol University Press.

Haskins, Caroline, Jason Koebler, and Emanuel Maiberg. 2019. "AirPods Are a Tragedy." *Vice* (blog), May 6, 2019. https://www.vice.com/en_ca/article/neaz3d/airpods-are-a-tragedy.

He, Megan, Jenna C. Ditto, Lexie Gardner, Jo Machesky, Tori N. Hass-Mitchell, Christina Chen, and Peeyush Khare, et al. 2024. "Total Organic Carbon Measurements Reveal Major Gaps in Petrochemical Emissions Reporting." *Science* 383, no. 6681: 426–32. https://doi.org/10.1126/science.adj6233.

Jang, Yong-Chul 2010. "Waste Electrical and Electronic Equipment (WEEE) Management in Korea: Generation, Collection, and Recycling Systems." *Journal of Material Cycles and Waste Management* 12, no. 4 (November): 283–94. https://doi.org/10.1007/s10163-010-0298-5.

Japan, METI Ministry of Economy, Trade and Industry. 2007. "Law for the Promotion of Effective Utilization of Resources." January 25, 2007. https://www.meti.go.jp/policy/recycle/main/english/law/promotion.html.

Japan, METI Ministry of Economy, Trade and Industry. 2020. "Circular Economy Vision 2020." https://www.meti.go.jp/shingikai/energy_environment/junkai_keizai/pdf/20200522_03.pdf.

Lane, Ruth. 2011. "The Waste Commons in an Emerging Resource Recovery Waste Regime: Contesting Property and Value in Melbourne's Hard Rubbish Collections." *Geographical Research* 49, no. 4: 395–407. https://doi.org/10.1111/j.1745-5871.2011.00704.x.

Liboiron, Max. 2013. "The Politics of Measurement: Per Capita Waste and Previous Sewage Contamination." *Discard Studies* (blog), April 22, 2013. http://discardstudies.com/2013/04/22/the-politics-of-measurement-per-capita-waste-and-previous-sewage-contamination/.

Liboiron, Max. 2014. "Against Awareness, for Scale: Garbage Is Infrastructure, Not Behavior." *Discard Studies* (blog), January 23, 2014. http://discardstudies.com/2014/01/23/against-awareness-for-scale-garbage-is-infrastructure-not-behavior/.

London, Bernard. 1932. "Ending the Depression through Planned Obsolescence."

MacBride, Samantha. 2012. *Recycling Reconsidered: The Present Failure and Future Promise of Environmental Action in the United States*. Cambridge, MA: MIT Press.

MacBride, Samantha. 2022. "Waste Metrics from the Ground Up." In *The Routledge Handbook of Waste Studies*, edited by Zsuzsa Gille and Josh Lepawsky, 169–95. New York: Routledge.

Ministry of Environmental Protection of the People's Republic of China. 2017. "World Trade Organization, Committee on Technical Barriers to Trade." World Trade Organization.

Minter, Adam. 2020. "China Finally Makes Its Peace with 'Foreign Garbage.'" Bloomberg.com, July 30, 2020. https://www.bloomberg.com/opinion/articles/2020-07-30/china-finally-makes-its-peace-with-foreign-garbage.

Packard, Vance. 1960. *The Waste Makers*. New York: D. McKay Co.

Patel, Nilay. 2017. "Show Notes: Cars Are Just Rolling Computers." The Verge, March 13, 2017. https://www.theverge.com/2017/3/13/14914784/show-notes-cars-are-just-rolling-computers.

Reed, John, Leslie Hook, and David Blood. 2018. "Why the World's Recycling System Stopped Working." *Financial Times*, October 25, 2018. https://www.ft.com/content/360e2524-d71a-11e8-a854-33d6f82e62f8.

Schulz, Yvan, and Benjamin Steuer. 2017. "Dealing with Discarded E-Devices." In *Routledge Handbook of Environmental Policy in China*, edited by Eva Sternfeld, 1st ed., 314–28. Abingdon, Oxon; New York: Routledge. https://doi.org/10.4324/9781315736761-27.

Singapore National Environment Agency. 2019. "Extended Producer Responsibility (EPR) for E-Waste by 2021." https://www.nea.gov.sg/docs/default-source/media-files/news-releases-docs/cos-2019/cos2019-e-waste-management.pdf.

Singapore National Environment Agency. 2021. "National Voluntary Partnership for E-Waste Recycling." https://www.nea.gov.sg/programmes-grants/schemes/national-voluntary-partnership-for-e-waste-recycling.

Slade, Giles. 2006. *Made to Break: Technology and Obsolescence in America*. Cambridge, MA: Harvard University Press.

Strasser, Susan. 1999. *Waste and Want: A Social History of Trash*. New York: Metropolitan Books.

Taffel, Sy. 2022. "AirPods and the Earth: Digital Technologies, Planned Obsolescence and the Capitalocene." *Environment and Planning E: Nature and Space*, January 2022. https://doi.org/10.1177/25148486221076136.

Terazono, Atsushi. 2013. "Recycling Small Home Appliances and Metals in Japan." July 2013. https://www-cycle.nies.go.jp/eng/column/page/202005_01.html.

United States Environmental Protection Agency. 2011. "National Strategy for Electronics Stewardship (NSES)." Overviews and Factsheets, July 20, 2011. https://www.epa.gov/smm-electronics/national-strategy-electronics-stewardship-nses.

United States Geological Survey. 2022. "2017–2018 Minerals Yearbook: Mexico." https://pubs.usgs.gov/myb/vol3/2017-18/myb3-2017-18-mexico.pdf.

United States Geological Survey. 2023. "2019 Minerals Yearbook: Republic of Korea." United States Geological Survey. https://pubs.usgs.gov/myb/vol3/2019/myb3-2019-south-korea.pdf.

Volk, Tyler. 2004. "Gaia Is Life in a Wasteland of By-Products." In *Scientists Debate Gaia: The Next Century*, edited by S. Schneider, J. Miller, E. Crist, and P. Boston, 26–36. Cambridge, MA: MIT Press.

Wynne, Brian. 1987. *Risk Management and Hazardous Waste: Implementation and the Dialectics of Credibility*. London: Springer London, Limited.

Xun, Sean. 2018. "The Mineral Industry of China." United States Geological Survey.

Yun, Jun-sik, and In-sung Park. 2007. "Act for Resource Recycling of Electrical and Electronic Equipment and Vehicles." Eco-Frontier. http://www.env.go.jp/en/recycle/asian_net/Country_Information/Law_N_Regulation/Korea/Korea_RoHS_ELV_April_2007_EcoFrontier.pdf.

Zimring, Carl A. 2011. "The Complex Environmental Legacy of the Automobile Shredder." *Technology and Culture* 52, no. 3: 523–47. https://doi.org/10.1353/tech.2011.0117.

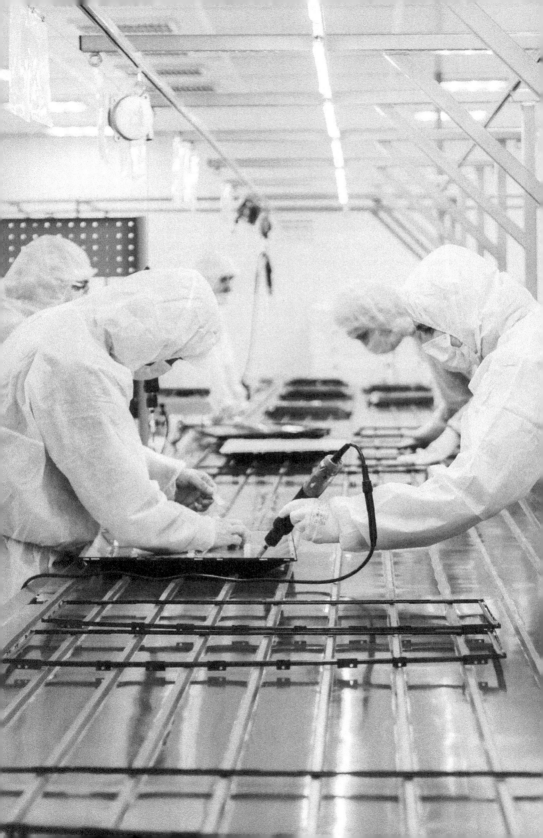

2

Problems, Controversies, and Solutions

Current global trends mean consumers will be discarding more than 100 million tons of e-waste annually between 2030 and 2040. Yet, far more pollution and waste arise from the mining for, and the manufacturing of, electronics. One of the most persistent problems associated with public understandings of e-waste is the nearly automatic association of the issue with what consumers get rid of, but with far less attention to problems of pollution and waste arising before they even purchase their devices. How a problem is framed is itself important to consider because a given framing makes some possible solutions seem obvious and takes other possibilities off the table while making still other potential solutions unimaginable in the first place.

This chapter provides an overview of some of the most persistent conceptual and empirical problems and controversies characterizing the issue of e-waste. It then turns to an overview of potential solutions for eliminating and/or mitigating pollution and waste arising from electronics throughout their existence, including from the mining of materials for them, their manufacture, their use, and beyond.

The End-of-Pipe Problem

The problem of e-waste began to be articulated in the early 2000s (see Chapter 1). Since its earliest days, the issue has been framed as what to do about electronics

---

After mining, manufacturing is the second most important source of pollution and waste arising from electronics. Here people work on an assembly line manufacturing flatscreen televisions. The clothing they wear is partially about their safety, but is mostly about protecting the electronics from contamination from human hair and dust that can harm the functioning of electronics. Manufacturing electronics is a chemically intensive process and leads to the release of toxicants and potent greenhouse gases. (Fotoeventstock/Dreamstime.com)

discarded by consumers and households. Data are collected and statistics collated about the growing mass of such discarded electronics. Legislation is passed to regulate the collection and processing of these discards, construed as e-waste. There is nothing wrong with defining the waste arising from discarded electronic devices as "e-waste," but framing the issue in terms of e-waste primarily or exclusively arising from discarded electronics has very important consequences. It is crucial to be mindful of those consequences if for no other reason than they make certain kinds of solutions seemingly obvious (e.g., recycling), while simultaneously making other solutions imperceptible even if they might reduce or eliminate pollution and waste arising from electronics in more substantive ways.

Presupposing that e-waste is that which arises after consumers or households get rid of their devices can be understood as framing the issue as an end-of-pipe problem. The "pipe" in this case is a metaphor. Obviously, finished electronic devices ready to be purchased by consumers do not literally flow out of pipes— although the fossil carbon needed to make some of their components may, indeed, have flowed through actual pipes at some point. As a metaphor, what "end-of-pipe" is meant to highlight is that the waste and pollution arising after consumers or households get rid of their devices are only one step in the overall throughput of materials and energy that are involved in the mining for, and manufacturing of, electronics long before consumers ever purchase their finished devices.

The concept of "throughput" accords with the end-of-pipe metaphor in a way that helps one think about how e-waste is shaped or framed as a problem to be solved. Throughput is about flow (e.g., of materials and energy). "Flow" suggests a system, like a river system, in which some parts of the system are "upstream" (e.g., at the headwaters of a river) and some parts of the system are "downstream" (e.g., an estuary or a flood plain). "Flow," "upstream," and "downstream" are additional metaphors that help one to understand how the shaping or framing of a problem like e-waste matters. It is crucial to pay attention to the work that the shaping or framing of a problem does to format which interventions or solutions seem obvious and are considered, which ones are dismissed, and which ones may not even enter the conversation at all because of the way the problem has been framed in the first place.

Again, being concerned about what is "emitted" at the end-of-the-pipe is not wrong or misguided. It is, however, a very particular way of being concerned about pollution and waste arising from electronics; and it is a way of being concerned that has important consequences. One such consequence is that concerns are directed to the smallest portion of the overall pollution and waste arising from electronics, as most waste is produced in the earlier mining and manufacturing stages of the pipeline.

Electronics are made from multiple materials including everything from plastics, to metals, to different types of glasses. One important material for making electronics is copper. The electronics sector is the largest consumer of copper after the construction industry, according to the United States Geological Survey (USGS). Copper is mined around the world, but one of the biggest such mines is the Chuquicamata Copper mine in Chile. A USGS study estimates that for every kilogram of copper mined, more than 325 kg of mine waste arise (United States Geological Survey and Goonan 2005). The same study shows that a minimum of 109,000 kt of mine waste arises annually at the Chuquicamata mine. Table 2.1 shows different categories of material flows associated with major copper smelting operations in Chile. Categories under "recycling loads"—revert, dust, matte—are each different mixtures of materials resulting from primary smelting that can be fed back into the smelting process for additional rounds of material extraction. Note the magnitude of outputs such as mine waste, mill tailings, slag, and other by-products. The mass of material under the mine waste category at a single site such as Chuquicamata (109,000 kt/yr or 109,000,000 tons/year), for example, is more than 1.7 times larger than all post-consumer e-waste arising on Earth in 2022 (62,000,000 tons; see also data in Table 1.1).

Seeing the difference between the mass of useful metal (1 kg) and the mass of waste arising (325 kg) might seem surprising, but it is typical in the mining industry where 98 percent or more of the total mass of material moved is considered waste by the industry itself (Keeling 2012). A skeptical reader might also ask whether this mass of material is really waste because it was already present at the site and has simply been moved from the subsurface to the surface during the mining process. A response to that skeptic might point out that far more is going on than the mere movement of subsurface materials to the surface. Mining waste in general, and copper mining waste specifically, is typically associated with the release of various toxicants, heavy metals, and acidic leachates due to previously subsurface materials being exposed to environmental conditions at the surface (e.g., precipitation; wider temperature ranges, sunlight, etc.). Water is especially important in these processes since it can pick up and move toxicants, heavy metals, and acids that bring harms to the ecologies through which the water moves (Piatek et al. 2007).

Table 2.1 Annual Material Flows Associated with Copper Smelting in Chile (thousand metric tons per year, kt/yr)

Flows	Altonorte	Caletones	Chagres	Chuquicamata	Las Ventanas	Potrerillos	Total
Recycling loads							
Revert	14	92	21	170	42	57	395
Dust	57	11	28	32	35	47	208
Matte	26	121	33	63	33	99	375
Outputs							
Mine waste	81,500	0	21,600	109,000	24,600	18,600	255,000
Mill tailings	22,500	37,600	12,300	42,300	10,200	22,000	147,000
Slag	311	696	188	607	288	275	2,360
Sulfuric acid	644	721	313	869	357	319	3,220
Sulfur dioxide	32	144	14	106	3	109	407
Carbon dioxide	38	18	6	87	25	8	183
Anode copper	267	368	138	460	138	202	1,570

Source: United States Geological Survey and Goonan (2005).

Figure 2.1 shows the mass of waste arising at a single copper mine relative to the mass of waste arising in three other situations relevant for thinking about pollution and waste from electronics: manufacturing (Eurostat 2019a), transboundary flows (Basel Secretariat 2021), and post-consumer discard (Eurostat 2019b). The figure shows the relative magnitude of annual waste arising in kilotons (kt) at the Chuquicamata mine in Chile (left) and electronics manufacturing facilities in the EU (center left), compared to global estimates of transboundary flows (center right) and "end-of-pipe" household e-waste collected in the EU (right). The figure also shows what a 1 percent reduction in waste at "upstream" phases (mining and manufacturing) would be equivalent to at the "downstream" phase of post-consumer e-waste management. Figure 2.1 makes it clear that the mass of waste arising at the end-of-the-pipe—that is, from post-consumer discard—is the smallest part of the overall issue. Framing the e-waste problem as an end-of-pipe problem is clearly limiting in regard to imagining and implementing effective solutions, as nearly all regulatory concern and consumer advocacy around e-waste have been directed toward addressing discarded electronics. Although this part of the problem is not unimportant, prioritizing the relatively small consumer waste portion of the issue has left the largest contributors to the overall problem of e-waste untouched. Understanding this point demonstrates why it is important to think critically about how the e-waste problem is framed as a problem in the first place.

Chapter 1 described legislation in various jurisdictions around the world that have been enacted since the 1990s to mitigate e-waste. Almost all of that legislation is directed to handling post-consumer discarded electronics as if what comes out of the end of the pipe constitutes the whole problem. Even if all of that legislation was completely effective and 100 percent of all the post-consumer discarded electronics were cared for as those laws state they should be, the result would never add up to the mass of waste arising from manufacturing electronics or from mining the materials needed to make them. Indeed, Figure 2.1 demonstrates that finding a way to reduce the mass of waste arising at a single copper mine by 1 percent would be the same as finding a way to reduce the waste arising from post-consumer discarded electronics by about 28 percent across the *entire* EU region.

There are other ways that the shaping of the e-waste problem matters. For example, formatting the problem in terms of mass can make certain kinds of comparisons possible, like those shown in Figure 2.1. But it is important to remember that electronics are composed of multiple materials, a number of which

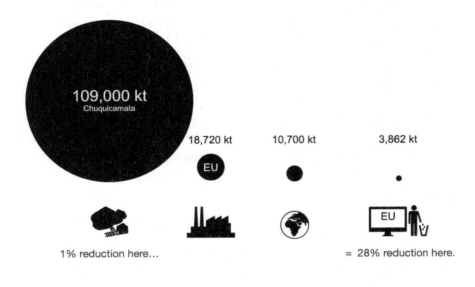

Figure 2.1 Relative Magnitude of Waste Arising over the Lifecycle of Electronics
Source: Modified from Lepawsky (2020).

can be toxic. There are also substances used in the manufacturing of electronics that are toxic but that are not incorporated into devices themselves. There is no simple relationship between mass and toxicity. Consider, for example, how 1 g of copper and 1 g of mercury—both of which occur in electronic devices—have the same mass but are completely different in terms of toxicity. Once again, it is not incorrect to specify the e-waste problem in terms of mass or of toxicity, but no matter what option is chosen, one is framing the problem in certain ways and not others. Framing the e-waste problem as one of mass might point one to solutions like household recycling. Framing the problem as one of chemical toxicants might lead one to proffer other solutions, like green chemistry in manufacturing.

There is a way in which the metaphor of "flow" built-in to the end-of-pipe framing of e-waste might be thought about critically so as to offer an alternative framing of the issue. Imagine arriving home one day to find the bathroom overflowing with water from the tub that someone forgot to turn off before leaving the house. One could rush to grab a mop and frantically run around the room trying to soak up the water flowing out of the tub. Doing so would not be an incorrect response, but it is probably not the best one. It would be much more effective to first turn off the tap out of which the water was flowing and then get to the mopping. Mopping up the water is a downstream solution, like consumer-based repair or post-consumer recycling. Turning off the tap is an upstream solution, like reducing the waste arising in the mining for, and manufacturing of, electronics in the first place. Framed this way the e-waste issue is a stock-and-flow problem, rather than an end-of-pipe problem (Liboiron 2015). Stocks, in this case, would be pollution and waste measured in terms of mass or toxicity. Flows, on the other hand, are the movement of materials or energy through the system and could, likewise, be measured in terms of mass or toxicity. The point here is to understand how problem framing affects how solutions may or may not be imagined and implemented. Some sort of framing is inevitable. The important thing to keep in mind is to always think critically about the consequences of framing e-waste and other pollution problems like it in particular ways and not others. The issue here is less about "right" versus "wrong" and more about how well or badly a pollution problem like e-waste is framed with respect to the solutions that might be offered to mitigate or eliminate it. It is important to test whether a given solution put forward to solve a problem like e-waste matches up with the magnitude of the problem the solution is claimed to fix.

Classification and Regulatory Definitions of E-waste

Classifying waste in general and electronic waste specifically is surprisingly tricky. An example of this in the previous chapter made this point: both the United Kingdom and the United States regulate the concentration of PCBs, which are known to be toxic. Yet, the regulatory threshold for PCBs in landfills in the United States is five times higher than that for landfills in the United Kingdom. The science of toxicology can create trustworthy knowledge that demonstrates PCBs to be toxic, but toxicology cannot adjudicate the US regulatory threshold for PCBs as "wrong" and the UK's as "right." Yet, policymakers often draft the sciences into such political situations under the false assumption that the results of science will provide policymakers with value-neutral results from which decisions can be made. One reason policymakers do so is that it seems to offer them a way to make decisions that are merely "technical" rather than political. Yet, no such easy and clear separation between value judgments and decisions exists. Conditions like this led Science and Technology Studies scholar Brian Wynne to argue that "wastes exist in a twilight zone where no clear, 'natural' definition of them can be given, within wide margins of uncertainty and variation" (Wynne 1987, 1). Wynne rejects the prevailing policy assumption that "the key terms 'hazard' and 'waste' can *even in principle* be precisely defined" (Wynne 1987, 8, emphasis in the original). The issues of defining "hazard" and "waste" detailed in Wynne's work are concretely evidenced, for example, in the still unsettled debates over the meaning of waste versus non-waste electronics under the provisions of the Basel Convention that have been ongoing since 2010, as discussed in Chapter 1.

Wynne argues that in situations where hazard and waste are being defined one is confronted not just with uncertainty, but with indeterminacy. The distinction between uncertainty and indeterminacy is subtle but important. Uncertainty describes a situation in which the behavior of a system can be known within a certain range. The present state of the system may not be knowable, but it is possible to know with certainty that the system cannot exceed state *A* or fall short of state *B*. An indeterminate system is fundamentally different. In an indeterminant scenario, even if it is possible to measure the current state of a system, it is impossible to know what range of values or behaviors it can move through.

An example of where these differences between uncertainty and indeterminacy matter can be found in a real situation of pollution arising from the manufacture of electronics. Trichloroethylene (TCE) is a chemical compound with a long history of use in manufacturing electronics, including in the region now dubbed Silicon Valley (Marisa Elena Duarte [Pascua Yaqui] and Jacob Meders [Mechoodpa/Maidu] 2021; United States Environmental Protection Agency 1989). Historically, TCE was disposed of from electronics manufacturing facilities in Silicon Valley through building drains that eventually led to the region's sewer system. TCE and other chemicals in the mix are acidic and corroded the system's pipes, which eventually leaked and formed a plume of toxic ground water in the region. After workers and broader communities in the region advocated for the need to clean up this toxic plume threatening their health, the US Environmental Protection Agency (EPA) designated the plume a "Superfund" site (essentially making it eligible to receive public money to clean up toxic spills from private businesses).

Problems, Controversies, and Solutions 47

As part of this Superfund cleanup process, which is still ongoing, the EPA used a mathematical formula to calculate the risk of harmful exposure from TCE to workers and residents in the region. TCE is a known carcinogen and can negatively affect the health of various human organs. The mathematical calculation used by the EPA is called the lifetime average daily dose (LADD) (United States Environmental Protection Agency 2010). The LADD formula makes certain assumptions to define what counts as a lifetime and what counts as an average daily dose. For example, the equation assumes that a person weighs 70 kg, drinks 2 L of water a day, and is exposed to TCE at a work facility or at home, but not both. All these assumptions are consequential and lead to a situation in which indeterminacy, rather than uncertainty, is the rule of the day. Note, for instance, that the definition of "person" in this formula would exclude anyone who weighs less than 70 kg. Most children would be excluded by these criteria even if they live in a home subject to TCE pollution from the plume of ground water under it. Similarly, the formula does not accurately account for the exposure of a person who both works at a facility where TCE exposure occurs and lives in a home that is also subject to TCE exposure because it sits on top of the pollution plume. Several other consequences of the formula's framing of LADD follow from it. For example, the formula offers no way to connect exposure in the workplace and in the home, it does not account for the possible toxicological consequences resulting from combinations of TCE with other chemicals, nor does it account for cumulative effects. A consequence of the overall situation is that there are significant numbers of differently situated people who can never know their LADD for sure. Their personal LADD is indeterminate, rather than uncertain, in this sense.

Debates about how to classify e-waste have, for the most part, followed the false hope of finding value-neutral, ever more precise technical definitions of electronic waste to overcome disagreements about what should count for the purposes of regulation. Recall the discussion of e-waste legislation in different jurisdictions in Chapter 1 of this book. The legislation in each country and region differs in many individual details. However, they all share a need to classify what will count as waste electronics. One or some combination of three approaches to classification is typically used: categories of devices, size of devices or a key component such as a screen, and voltage. One consequence of taking these different paths to classifying waste electronics is that the definitions in different jurisdictions wind up being incommensurable with one another. For example, even when two different jurisdictions such as the

EU and South Korea incorporate voltage as a key characteristic for defining e-waste, South Korea explicitly includes electric vehicles while the EU explicitly excludes them. These differences in classification decisions have no non-arbitrary basis yet can lead to very different measurement outcomes for e-waste arising in the two jurisdictions since electric cars are much more massive than, say, laptops or desktop computers.

Indeterminacy and incommensurability are two inescapable challenges that come with any attempt to classify e-waste. However, just because these challenges are always present, that does not mean it is impossible to manage e-waste well. Instead, indeterminacy and incommensurability need to be acknowledged and, in so doing, shift debates about how to regulate e-waste toward assessing how well or how poorly this or that classification of e-waste works for achieving this or that desired outcome for mitigating or eliminating pollution and waste arising from electronics. All too often however, these challenges of indeterminacy and incommensurability with respect to e-waste are skipped over as policymakers rush to implement seemingly obvious solutions, such as consumer right-to-repair or post-consumer recycling. Yet, such solutions are only obvious in situations where the underlying, unstated assumptions take as an uncritical truism that e-waste is what consumers or households get rid of when they toss their devices. Given how much pollution and waste arise during the mining for, and the manufacturing of, electronics no amount of post-consumer recycling of discarded electronics will make up for them. Although necessary, neither consumer-based reuse, nor repair, nor recycling is sufficient to mitigate or eliminate the magnitude of challenges posed by pollution and waste arising from the electronics sector. That does not mean that reuse, repair, and recycling of e-waste classified as that which arises when consumers discard their devices should not be done, but it does mean that such a classification of e-waste is insufficient to substantively mitigate—let alone eliminate—total pollution and waste arising from electronics.

Magnitude: How Big a Problem Is E-waste?

Most discussions of e-waste begin with an accounting of its magnitude. Magnitude is the size or extent of something, but the word also connotes importance. Things of larger magnitude are typically understood to be more important than things of smaller magnitude. For example, according to the

Problems, Controversies, and Solutions

global e-waste monitor 62 Mt of post-consumer electronic waste were generated globally in 2022 (Baldé et al. 2024). As noted in Chapter 1, this figure represents a growth rate of about 5 percent per year since 2019. Again, such a growth rate may appear small, but it represents a doubling time of about twenty years. That means that by around the year 2040 approximately 100 Mt of post-consumer electronic waste will arise globally if current trends continue unchanged.

The idea of 100 million metric tons of post-consumer electronics piling up twenty-five years from now might seem like a large magnitude problem. But it is important to ask: large compared to what? Recall that mining for, and manufacturing of, electronics generates substantially more pollution and waste than do consumers when they discard their devices. Indeed, as Figure 2.1 shows, the copper mine at Chuquicamata, Chile, already generates about the same mass of waste annually 109,000 kt—or 109 million tons—that projections of total post-consumer e-waste suggest will arise globally somewhere around 2040.

Meanwhile, data for releases of chemical toxicants from the electronics manufacturing sector offer another perspective on the question of magnitude. Data available for Canada, Mexico, and the United States show that between 2006 and 2020 the electronics manufacturing sector released a total of 447 million kg (or 447,000 tons) of chemical toxicants. These releases of toxicants are notable because they represent a mass of materials that are not incorporated into electronic devices themselves. In other words, they represent a mass of materials that thinking about e-waste in strictly post-consumer terms misses all together.

The prospect of 447 million kg of hazardous chemical releasing over a period of fourteen years in three countries is, of course, of lower magnitude than the 62 million tons of post-consumer e-waste arising globally in 2022 (Baldé et al. 2024). Yet, magnitude needs to be thought carefully about here as well. The figure for 447 million kg of chemical toxicant releases comes from a database maintained by the Commission for Environmental Cooperation (CEC), an organization under the Canada-US-Mexico Agreement (CUSMA, formerly NAFTA). This database is part of a general category of regulatory data repositories called pollution release and transfer registries (PRTRs). NAFTA (now CUSMA) was the first international trade agreement to require the maintenance of a PRTR. The CEC requires all industries operating in the NAFTA/CUSMA region to report the release of nearly 700 different chemicals deemed hazardous. Data for the electronics manufacturing sector are available from the CEC database. Those data show that the 447 million kg released from the sector are comprised of 104 distinct chemical toxicants. However, as of 2024 there are more than 313 million chemicals available for industrial use of which only 196,000 have ever been tested, and of those, all have been found to be toxic. What do all these numbers mean with respect to the question of the magnitude of the e-waste problem?

Here are some ways to think about answers to this question. The CEC data show that the electronics sector is responsible for the release of 104 different hazardous chemicals. But, the 700 or so different chemical releases mandated to be recorded by industry in the CEC database is six orders of magnitude less than the 313 million chemicals currently available for industrial processes. That means that only a relatively tiny number of chemical toxicants (~700) are even required to have data collected about their release from manufacturing facilities in the CUSMA region. Consequently, knowledge about chemical toxicants released by industry in general, and the electronics manufacturing sector specifically, is extremely limited. Not only is the figure of 447 million kg of chemical releases from the electronics sector a significant under count, the six orders of magnitude difference between the number of chemicals available for industrial uses (~313 million) versus the number of chemicals required to be reported (~700) means that the lack of information is so vast that the idea of a knowledge "gap" makes little meaningful sense. The following analogy may be helpful for understanding this point.

If the 196,000 chemicals tested and also found toxic were kilometers, they would stretch about half the distance between the Earth and the Moon. In contrast, if the 313 million chemicals currently available for industrial use were kilometers, they would stretch from Earth to slightly past the Sun. The huge relative difference between the number of chemicals tested and known to be toxic compared to all the chemicals available for industrial use is far too big to be described as a mere "gap" in toxicological knowledge. Indeed, the number of chemicals available for industrial use vastly exceeds all toxicological testing capacity on Earth. The only thing that is almost certain under such circumstances is that humans will never have full knowledge about the actual or potential chemical hazards that arise from the chemicals they have made available for industrial uses. Consequently, pollution and waste arising from the electronics sector generate a situation of indeterminacy rather than uncertainty, as discussed earlier. Noting this situation of indeterminacy, however, is not a reason to avoid devising solutions to the problems of pollution and waste arising from electronics. Various solutions are discussed later in this chapter.

Scale: Matching Problems with Solutions

In everyday language "scale" is often used interchangeably with the idea of differences of size. Yet, when it comes to questions of how to mitigate or eliminate the pollution and waste arising from electronics there are good reasons to think of scale as relationships that matter in which differences of size may be of secondary importance or even unimportant altogether. Thinking of relationships that matter helps to assess whether a proposed solution to a pollution or a waste problem is a good match for the problem so framed. Problem framing was discussed earlier. It was noted, for example, that two common ways of thinking about e-waste frame the problem space in which it is considered. One of those ways of framing the problem is presupposing e-waste to be an issue of post-consumer discard of devices. That way of thinking about e-waste almost always occurs in conjunction with a second framing of the problem as one that can be solved by implementing post-consumer recycling. Sometimes it also includes discussions of consumer reuse and repair. However, as previously discussed, no amount of consumer reuse, repair, or post-consumer recycling—even if they achieve a 100 percent success rate—can make up for the pollution and waste arising from the mining for, and manufacturing of, electronics (see Figure 2.1). This misalignment between the magnitude of waste arising in the mining for, and manufacturing of, electronics does not mean that post-consumer recycling is useless or should not be pursued. Such recycling can achieve desirable ends such as diverting materials from landfill and incineration, employment, and materials recovery for reuse in manufacturing. However, the misalignment does mean that post-consumer solutions like recycling cannot, on their own, mitigate or eliminate the much larger amount of pollution and waste occurring upstream of consumers who use electronic devices. Researchers have coined a term for this kind of misalignment between the framing of a pollution and waste problem and the solution(s) proposed to alleviate them: scalar mismatch (Liboiron 2014).

Reconsider the scene with the overflowing bathtub described earlier to better understand the implications of scalar mismatch. The flooded bathroom did not occur all by itself. Nor did all the many molecules of H_2O coming out of the tap automatically add up to a flood on their own. For that to happen, someone had to forget the tub was running and leave the house. Upon arriving home and discovering the flooding bathroom, it is true that one could grab a mop and vigorously try to soak up the water on the floor while the tap keeps running and the bathtub continues to overflow. Yet, the difference in size between water molecules, the mop, the bathtub and so forth is not really the salient issue here. It is the relationships between all these elements and the order of operations that matter. Grabbing a mop (while the tap keeps running) to achieve the cleanup is a misalignment between the problem at hand and the solution proffered (mopping). That is scalar mismatch. A better solution to the problem would be to turn off the tap and then begin mopping up. The latter approach appropriately aligns the solution with the problem.

All too often, advocacy around e-waste defaults to offering recycling as the solution to the problem. However, when all pollution and waste arising from electronics are considered, it is important to distinguish the relative contribution of different phases of the lives of electronics. Post-consumer e-waste is one phase, but it is different than the pollution and waste arising in other phases of the lives of electronics, such as mining and manufacturing. It is not only that each of these phases represents masses of pollution and waste that differ by orders of magnitude, but that they are different in their composition as well as having different sources and different responsible parties. Consequently, different forms of advocacy for change are needed that align with each of these differences. It is quite a different proposition, for example, to advocate for reduced pollution and waste arising from mining than it is to advocate for a municipal recycling system. Again, size is not really the issue here. Some mines are as big as municipalities, but the systems of accountability for mines are quite different than those from municipalities. Similarly, advocating to reduce or eliminate the use of toxic chemicals in the manufacturing of electronics must direct action toward different leverage points for change than does advocacy for post-consumer recycling. The issue is not whether one form of advocacy is right and another is wrong—arguably all forms of such advocacy are needed—but whether a given solution advocated for actually lines up with the pollution and waste problem it claims to be solving. A variety of solutions to the e-waste problem as a whole are discussed later in this chapter.

Jevons Paradox, Efficiency, and Rebound

One way that actors concerned about pollution and waste arising from electronics suggest solutions may be found is through improvements to efficiency. In this context, efficiency refers to a diverse set of strategies that could reduce

Problems, Controversies, and Solutions 53

the throughput of materials and energy in the making and use of electronic devices. Hardware and software engineers often focus attention on methods to improve efficiency in this sense. However, as a potential solution to pollution and waste problems arising from electronics, improvements in efficiency have important limits to what they can achieve. These limits have been known since the nineteenth century.

William Stanley Jevons was a nineteenth-century political economist who studied the use of coal in the burgeoning Industrial Revolution. His research led to a surprising finding: as coal using machines became more efficient—that is, they used less coal per unit of work achieved or product produced—aggregate demand for coal rose rather than declined. In other words, instead of improvements of efficiency leading to reduced demand for energy resources like coal, as might be expected, demand went up. At first, such a result may seem counterintuitive. Yet, some basic principles about relationships between price, supply, and demand help make sense of this seemingly surprising relationship. What Jevons recognized was the important scalar mismatch between construing per-unit efficiency improvements as a way to achieve aggregate reductions in demand. Improvements in per-unit efficiency mean that less of a resource like coal is needed to achieve the same amount of work as was previously necessary before the efficiencies were realized. But those efficiency gains have a knock-on consequence: they mean that the supply of a resource such as coal relative to the demand for it initially increases as efficiency improves. When supply is in excess of demand for something, prices for it typically fall, that is, it becomes cheaper to use this or that resource. In that circumstance, then, those who used coal before the efficiency gains can afford to use more. Meanwhile, those who may not have been able to afford coal at all before the efficiency gains can also now afford to do so. Consequently, overall demand for coal goes up. Subsequent economists named this relation after William Stanley Jevons, dubbing it the Jevons Paradox (it is also sometimes more technically referred to as the rebound effect).

What Jevons witnessed with coal can also happen with the materials and energy needed to manufacture and use electronics. For example, some brands of consumer electronics tout their use of recycled materials in the making of their devices. Some of those common materials include metals such as aluminum (Calma 2019). The use of recycled materials such as aluminum in manufacturing can lead to reduced energy and material throughput of this or that brand of electronic device maker. Yet at least two issues ensue that may undermine the hope that the use of such recycled materials in manufacturing will automatically lead to an overall reduction in materials and energy throughput. First, if one brand can reduce its use of primary materials by using secondary materials, such as recycled aluminum, the supply of primary aluminum relative to the number of customers goes up. Consequently, prices for that primary aluminum may fall leading to other brand manufacturers being able to afford more than they otherwise have in the past. That situation can cancel out the efficiency gain from this or that brand's use of recycled aluminum. Second, if the sales of a given brand's devices increase by a percentage equal to or greater than the percentage of recycled aluminum used per device—perhaps because green-minded consumers notice its use of recycled aluminum—the efficiency gain will be wiped out due to increased sales. Both possibilities represent instances of the Jevons Paradox or rebound effect.

Similar complications can arise with energy use and electronics. If emissions from the energy needed to run computers-in-use are considered part of a total picture of pollution and waste arising from electronics, then accounting for those emissions is relevant. But it is also extremely difficult, if not impossible, to calculate precise relationships between energy use, computing, and various emissions, such as CO_2. This difficulty arises because actual locations of computer processing and power grids offer so much variation in the mix of power generation types (e.g., hydro, coal, wind, solar, gas) that the idea of an "average" does not make much sense (Pasek 2023).

There is some evidence that total energy use is declining relative to the increase in compute time at data centers (Masanet et al. 2020). What is not yet clear is whether this apparent decoupling of the relationship between compute time and energy use is a temporary phenomenon that might be overwhelmed as new computing applications become more and more prevalent, for example, machine learning, sometimes known as artificial intelligence (Woodruff et al. 2023). The compute time–energy use relationship is complicated further by the complexity of existing energy grids and how they may or may not change over time—for example, if there is a concerted shift away from fossil fuel-based energy generation toward greater use of renewables such as solar or wind. Moreover, there is the further complexity of how much of what kinds of work can be done remotely via electronic devices that in the past required substantial personal mobility, for example, from home to office. In a recent report, the International Energy Agency estimated that approximately a third of jobs in "advanced economies" can be done remotely without the need of fossil fuel-based travel (International

Energy Agency 2022, 7). If employees in such jobs worked remotely three days per week, that could avoid the use of approximately 500,000 barrels of oil per day. Like other calculations, these are estimates, but they are suggestive and point to important areas for future research around the energy aspects of Jevons Paradox and whether there may be realistic scenarios in which pollution and waste arising from compute time may avoid the rebound effect.

Decoupling

At a basic level decoupling refers to the separation of rates of economic growth from pollution and waste arising. Usually, decoupling is specified as reductions in energy use and/or CO_2 emissions versus growth in gross domestic product (GDP). A substantive research debate exists around the issues of whether and how digital technologies may or may not enable decoupling. Using the substitution of remote work for office-based commuting as a concrete example, key points of the debate involve at least two issues: (1) how the production, use, and disposal of digital devices themselves relate to aggregate material and energy throughput (e.g., whether a substitution of remote work for fossil fuel-based commuting leads to an overall drop in energy and material needs relative to economic activity), and (2) whether energy efficiency and productivity gains from digitization can achieve reductions in both per-unit and aggregate energy and material throughput (i.e., whether efficiency can overcome the Jevons Paradox or rebound effect).

Studies of decoupling at the national or international scale suggest that what evidence for decoupling does exist in actual economies would have to be radically more significant for their effects to have truly ameliorative consequences for CO_2-induced global heating (Jackson, Hickel, and Kallis 2024). For example, in a study of eleven high-income countries that have achieved a measurable degree of decoupling, researchers found that decoupling rates would need to increase ten times in order to reach those countries' responsibilities under the Paris climate accord (Vogel and Hickel 2023). In contrast, other researchers have shown empirically that improvements in human well-being do not inherently require continually increasing energy consumption or carbon emissions to be achieved. In other words, there are examples of countries that achieve high levels of human development at much lower levels of per capita energy consumption and carbon emissions than others with the same or similar levels of human development (Steinberger and Roberts 2010).

Meanwhile, studies of decoupling of the electronics sector offer a mixed picture. Several classes of electronic devices have been experiencing a reduction in the mass of materials composing them over time. This is a process sometimes called "lightweighting" in the industry (Linnell 2018; Staub 2017). Detailed analysis, however, shows that although devices are becoming less materially intensive per device, those material efficiency gains are being wiped out by increased aggregate consumption. For example, data from the United States show that between 1990 and 2010 overall consumption of electronic devices increased from 0.5 new products per year to 3.5 new products per year (Kasulaitis, Babbitt, and Krock 2019). That is an increase of 400 percent in the consumption of devices. The average mass of devices declined from 16 kg to 4 kg over the same period or a 400 percent decrease in average mass. So, while lightweighting is increasing the material efficiency of creating devices, increased consumption is exceeding those efficiency gains—just as the Jevons Paradox or rebound effect would predict. Currently then, it seems that the best that material efficiency gains can be said to achieve is that they slow, but do not fully decouple, material throughput from device manufacturing.

On the energy side of the decoupling question, there is evidence that between 2010 and 2018 decoupling has "substantially" and "mostly," if not completely, occurred between the growth in data centers and aggregate energy use (Masanet et al. 2020, 985). However, these findings pre-date the arrival of new energy-intensive data center applications such as large language learning models (LLMs or "artificial intelligence"). The energy footprint of LLMs has only recently begun to be studied. Results thus far suggest that these applications are highly energy intensive (Strubell, Ganesh, and McCallum 2019). Consequently, it remains unclear whether the substantial decoupling of energy use and growth of data centers is a temporary or long-term trend that will continue or, instead, be wiped out by growth in aggregate energy use for LLMs and similar applications as the Jevons Paradox or rebound effect would predict.

Part of the issue may be shifting notions of efficiency to sufficiency. It is common for economic and engineering models of systems to presume material abundance and energy scarcity. From a material point of view, the Earth system is effectively a closed system (significant new additions of material are added to the Earth system only through occasional and catastrophic collisions with other celestial bodies like comets). However, from an energy perspective the Earth system is open since it receives massive amounts of energy from the Sun. Indeed, calculations show that only 0.04 percent of incoming solar radiation would be required for all inhabitants on Earth to access as much energy per capita as a typical Canadian (Chachra 2023). Given that situation, new economic and engineering models could be reoriented to match the scenario that the Earth system actually finds itself in—that is, one of material scarcity, but energy abundance. Designing electronic devices from this new perspective could lead to a reorganization of the industrial systems that mine for and manufacture those devices as well as reorganizing how those devices are distributed, used, and reclaimed at the end of their useful lives. These issues relate to solutions to pollution and waste problems arising from electronics and are discussed in more detail later in this chapter.

Transboundary Flows

Transboundary flows of post-consumer e-waste remain controversial because of their association with waste dumping. In this context, waste dumping describes the actual or potential movement of post-consumer e-waste across borders to jurisdictions with weaker or more poorly enforced environmental legislation than those in jurisdictions from which shipments of e-waste originate. The explanation for why waste dumping occurs is that there is a cost savings to be had for an exporting jurisdiction by avoiding costs for proper e-waste management within its borders. The cost savings explanation for waste dumping holds true in theory. Yet, in practice the picture is much more mixed. There is also no question that actual examples of waste dumping have and do occur. Activism, journalism, and scholarly research have developed a robust evidence base documenting empirical examples of waste dumping across various types of borders (e.g., international borders, subnational boundaries such as state or provincial territories, and socioeconomic boundaries such as those that divide municipalities or regions along lines of race and class). However, when it comes to transboundary movement of post-consumer e-waste the explanatory picture is more complicated than can be conveyed in an overly broad category like "waste dumping." Classic examples of waste dumping have involved materials like incinerator ash, chemical hazardous waste, and municipal solid waste. Even in the narrow sense of post-consumer e-waste, discarded electronics typically have a more complex relationship with questions of waste and value than the other types of waste just mentioned. Some constituent materials of post-consumer e-waste, such as precious and semi-precious metals, can be lucrative to recover when prevailing market conditions are right. What is more often the case is that the reuse value of electronics discarded by consumers in relatively wealthy markets is far higher than the value of materials that could be recovered from those discarded devices, especially if they are moved across borders.

In relatively wealthy markets, it is not unusual for consumers to discard their devices in situations where those devices become available at little or no cost to other people who can refurbish and/or repair them and sell them back into reuse markets locally or, alternatively by shipping them across borders. For example, a report published in 2015 by a consortium of United Nations entities, Interpol (the International Police Organization), trade associations, and research consultants examined the issue of illegal transboundary shipments of e-waste associated with the EU (Huisman et al. 2015). This analysis found that "the main economic driver behind these shipments [of e-waste departing the EU] is reuse and repair and not the dumping of e-waste" (Huisman et al. 2015, 6). Actors engaged in exports of e-waste from the EU were, indeed, found to be avoiding some types of costs, but not those associated with proper disposal within the EU. Instead, the costs avoided were related to sorting, testing, and packaging of devices for reuse, not the costs of proper waste management within the EU. The reuse value of devices once exported was found to far exceed the costs of proper disposal within the EU and to exceed the value of materials that might be derived from recycling operations either within the EU or abroad.

If waste dumping is going to be used to explain why transboundary shipments of electronics discarded by others happen, then that explanation must account for the actual costs of shipping. When it comes to shipping costs it is helpful to consider an analogy that readers might already be familiar with—purchasing commodities online and having them shipped to one's home. In transactions like that, the purchaser pays for shipping, not the seller. The same is true in the international shipping world where containers move on ships and the commodities in them cross borders. The purchaser or receiver of the goods pays the cost of shipping, not the sender. That situation means that one must ask why a purchaser would pay to ship discarded electronics from one place just so they could be dumped in another place, if it costs more to ship them than to dispose of them properly in the location from where they are shipped.

There are cases where electronics recycling companies based in exporting countries (e.g., the United States) have been caught sending scrap electronics (i.e., electronics that are unusable and/or unrepairable) to buyers in other countries (Staub 2019). But companies that do this are not avoiding disposal costs. They receive discarded electronics either for free (e.g., through a community drop-off events), or from recycling programs that pay those companies for processing, or the companies charge customers to drop off unwanted devices. In any case, the recycling company has its costs covered. Receiving discarded electronics for free, via a recycling program, or charging for collection means that selling the devices for reuse or scrap value provides additional profit. So, it is not an attempt to avoid disposal costs that leads to transboundary shipments of e-waste in these cases. Instead, some recycling companies that market their services as "no export" are engaging in deceptive business practices as sellers of services they do not actually provide. They recycle devices and they export for additional profit, even as they advertise a "no export" approach. Such a practice is false advertising and misleads customers who may select a recycling firm in whole or in part because of its stated no export policy.

In contrast, a buyer who is willing to pay the costs of the discarded electronics being offered for sale and pay for the costs of shipping is doing so because they believe they will be able to recoup all those costs, plus make a little extra (profit) by selling those devices for reuse or scrap in the import market. The reality on the ground of such a transaction can be quite unsavory. An electronics recycling company that advertises itself as one that does not export, but that then turns around and sells to export markets is, at best, engaging in deceptive marketing. In some cases, companies have been found to be breaking the law. Meanwhile, the importing customer may find that the condition of discarded electronics that they purchased does not permit for repair or refurbishment for reuse, leaving only the value of materials in local scrap markets to be had. The latter may not be enough to turn a profit, in which case the importer loses money and also may abandon the scrap electronics imported. Situations like this do happen, but it is not the avoidance of disposal costs (i.e., waste dumping) that explains them.

When it comes to transboundary shipments of post-consumer e-waste, the issue is not that devices never reach an end to their usability at some point. They do. But when they do, new rounds of activity occur to recover components— such as capacitors, memory chips, and motherboards—that may be reusable. Materials recovery (recycling) also occurs, such as for plastics, metals, and glasses that can be processed for further rounds of value recovery, for example, to be sold as feedstock for manufacturers. It is true that no amount of recycling can completely recover all components and materials. Some amount of remainder

60 — *Electronic Waste*

(i.e., pollution and waste) will arise and it may be dealt with in ways that harm human and environmental health. However, this is a result that occurs with or without transboundary shipment. Indeed, there are studies that document the toxic burden on electronics recycling workers even in facilities that meet high standards of occupational health and safety (Beaucham et al. 2018; Ceballos and Dong 2016).

(Il)legality

The question of the legality or illegality of specific ways of managing e-waste depends on how e-waste is classified and in what jurisdiction the management takes place. As discussed earlier, such classification is challenging. The challenges of classification lead to unintended yet foreseeable clashes between what is considered legal in some jurisdictions, but illegal in others.

An illustrative example of the challenges of defining the line that divides the legal from the illegal is the ongoing saga of attempts to distinguish between waste and non-waste electronics under the provisions of the Basel Convention. Recall Figure 2.1, which shows that up to 10,700 kt of e-waste were thought to flow across borders globally in 2019. That figure comes from the 2020 Global E-Waste Monitor (Forti et al. 2020, 14) and is cited in a joint proposal by Ghana and Switzerland to amend various sections of the Basel Convention so that it could better manage waste electronics (Basel Secretariat 2021). Prior to the proposal by Ghana and Switzerland, the Basel Convention placed discarded electronic equipment, some of their components, and some of their materials on two separate lists, one defining waste and the other defining non-waste. The list defining waste, formally known as Annex VIII "List A," described equipment such as waste batteries, waste electrical and electronic assemblies, waste metal cables, and glass waste from cathode ray tubes (CRTs). Meanwhile, a separate list formally known as Annex IX "List B" defined non-waste, yet it also contains description of certain batteries, electronic assemblies, and waste cables all with only minor differences between those that appeared on List A. Consequently, prior to the proposal by Ghana and Switzerland, discarded electronics, some of their components, and some of their materials were both waste and non-waste under the meaning of the Basel Convention. This situation made for a confusing regulatory environment that was difficult at best to enforce. The proposal by Ghana and Switzerland was meant to mitigate these challenges in several ways. First, the proposal recommended adding a new category of waste to a different

Problems, Controversies, and Solutions

part of the convention formally known as Annex II "Categories of Wastes Requiring Special Consideration" (Basel Secretariat 2011, 55). The new category added to Annex II would cover all discarded waste electrical and electronic equipment, including scrap. Second, wording in Annex VIII List A would be changed to better reflect a comprehensive description of waste electrical and electronic equipment as well as its scrap. And third, entries in Annex IX List B that also described discarded electronic equipment, its components, and materials would be deleted.

The amendments suggested by Ghana and Switzerland were negotiated and adopted into the Basel Convention in 2022 (Basel Secretariat 2022). The amendments made important changes to the Convention that had previously categorized different fractions of e-waste as both waste and non-waste. The amendments also attempt to clarify language on permissible and non-permissible exports for activities such as reuse and repair of used electronic equipment (Basel Secretariat 2023b). However, the definition of "reuse" continues to "include repair, refurbishment or upgrading, but not major reassembly" without specifying what would count as "major reassembly" (Basel Secretariat 2021, 5). The definition of reuse, then, remains open to fairly wide interpretation. Moreover, the additional specifications of the meaning of reuse and repair under the Basel Convention are contained in a technical document that has only been adopted on an interim basis to date (Basel Secretariat 2023a). Consequently, after more than a decade of trying to define a bright line between legal and illegal transboundary shipments of e-waste that also contains provisions for legitimate reuse, repair, and recycling the current interim state of play means there still is no legally enforceable clear line between the legal and illegal that all parties of the Basel Convention agree to.

The high-level international negotiations at the Basel Convention have concrete implications for how legality and illegality are interpreted on the ground. For example, the study of transboundary shipments of e-waste out of the EU cited earlier found that about 1.5 million tons of discarded electronics were exported for reuse and repair. However, the study also found that of those 1.5 million tons, 1.3 million tons or about 86 percent exist in a "grey area subject to different legal interpretations and susceptible to export ban violations" (Huisman et al. 2015, 16). When over 80 percent of something that is supposed to be able to be distinguished as either legal or illegal cannot be so distinguished, it is a good signal that regulations need substantial improvement.

The challenge of drawing a line between legal and illegal action as it pertains to e-waste stems from a point made earlier: "waste" is a fundamentally ambiguous category (Wynne 1987). Legal theory discusses a concept called the work of jurisdiction which refers to the organization and conduct of legal authority over this or that question of legality (Valverde 2009). Decisions about legality and illegality are organized around questions like *what, who, when,* and *where* are legal decisions to have authority over. The order of the questions may be different in different legal cases but asking them leads to an answer of how legal authority is to work in a given situation. For the most part, e-waste legislation, such as the Basel Convention or the EU's legislation, begins with the issue of *what* is to be governed, that is, equipment, its components, and/or its materials that are deemed waste. Despite being a seemingly simple distinction between what will count as waste and what will not, wide ranging and ongoing controversies about defining that distinction carry on. Even within the examples of the Basel Convention and the EU's e-waste legislation, different versions of the question of what is to be legal or illegal are posed. For example, EU e-waste legislation is premised on definitions of waste that derive from what a person does, intends to do, or is required to do when discarding some thing or another (Jacobs 1997). In contrast, the definition of waste under the Basel Convention is based on the characteristics of materials being disposed of rather than the action of a person. This is an important difference to notice. Instead of trying to define the character of the objects being discarded as waste or non-waste, the EU legislation focuses on what a person is doing, will do, or must do with objects being discarded. There are many nuances to the legal regulation of e-waste in these examples, but these two different ways of defining what will count as e-waste demonstrate that different ways of organizing the distinction between legal and illegal behavior around e-waste are possible. Noticing such differences hint that there are a variety of alternatives available for dealing with these matters of legal concern. Such alternatives and solutions to e-waste problems are discussed later in this chapter.

E-waste, Environmental Racism, Environmental Justice, and Colonialism

Recall from earlier discussions in the book that the earliest concerns about discarded electronics emerged in the 1970s and were about how they might be used as a source of scrap precious and semi-precious metals. Only much later did the issue become about how brand electronics manufacturers might handle

Problems, Controversies, and Solutions

their recycling needs. But the issue of e-waste gained wider public attention after it was connected by ENGOs to concerns about environmental racism and environmental justice. Environmental racism and environmental justice are substantial fields of research in their own right. E-waste has become a charismatic case in these broader fields.

Both environmental racism and environmental justice are contested concepts that share connections but also important differences with one another. At a basic level, environmental racism is a way to describe patterns of toxic harm disproportionately impacting people racialized as minorities, impoverished, or both. In the United States, the origin of the term is often traced to public opposition to the dumping of polluted soil in Warren County, North Carolina, in the early 1980s (United Church of Christ Commission for Racial Justice 1987). Since then, the meaning of environmental racism has been broadened and nuanced through the notion of environmental justice. In contrast to the focus on harm baked into the concept of environmental racism, environmental justice may convey various ways of thinking about fairness and equity with respect to environmental conditions (Liboiron et al. 2023). Two prominent examples are distributional justice and procedural justice. Distributional justice conveys the idea that everyone—regardless of their social position in terms of race, class, or other social distinctions—has a right to a healthy environment. Procedural justice focuses on fair and equitable access to decision-making procedures related to environmental harms and benefits.

Anthropologists Reno and Halvorson (2022) offer a helpful way into the many complexities that link and mutually construct both waste and race. One of the key issues these authors identify is how people who happen to have a particular phenotypic characteristic—brown or black skin versus white skin for example—come to be associated with different forms of value and devaluation, including selective and hierarchical belonging to, and exclusion from, the category of "human" (Mitchell 2000, Chapter 9). Thus, Reno and Halvorson make the crucial point that the mutual construction of waste and race includes the manufacture of "blackness" and "whiteness." Whiteness, some have argued, is a racialization inhabited by people and places who are supplied with the power to be the default, taken for granted "norm" against which all else is automatically deemed and judged to be deviant and "Other." Said differently, whiteness is the power to accrue benefits that can come from being "invisible," while all else is marked out in some way for differential treatment as lesser (Benjamin 2019). As Reno and Halvorson point out through a discussion of the derogatory slur "white trash," the phenotypic characteristic of white skin is, in and of itself, no guarantee of being able to fully inhabit whiteness and accrue its benefits. Citing the work of historian Nancy Isenberg (2016), Reno and Halvorson (2022, 46) note how people who were poor and landless but happened to have white skin, "served as a cover for white elites deflecting attention from class inequality while promoting stories of rags-to-riches, individual (white) social mobility." Waste and race, then, mutually manufacture both whiteness and its Others, even as the harms and benefits of those racial categories are unevenly distributed and experienced by those who inhabit one form of racialization or another.

What Reno and Halvorson (2022) are arguing about waste and race more generally is also relevant to controversies about e-waste. The earliest ENGO reports on e-waste centered on a narrative of environmental racism premised on the notion of waste dumping—a prominent theme of some of the earliest publications in the broader environmental racism literature. Reports from the Basel Action Network (2002), Greenpeace (Greenpeace International 2008), and others (World Health Organization 2021) frequently used images of people racialized as non-white surrounded by heaps of what viewers were meant to assume were junk electronics from elsewhere. Journalists amplified the framing of environmental racism and waste dumping. These reports often pointed to the idea of transboundary shipments of e-waste from wealthier—and at least implicitly, "whiter"—markets as the key source for these piles of junk electronics. Later research by other ENGOs and some scholars raised questions about the veracity of such claims. For example, a report on the news program *Frontline* claimed that "hundreds of millions of tons" of e-waste were imported each year to, "one of the world's digital dumping grounds" called Agbogbloshie, a site in Accra, the capital of Ghana (Klein 2009). Even as some researchers accepted those claims, others pointed out that the amount of container traffic moving in and out of Ghanian ports would have to have been ten times higher than what publicly available data showed at the time (Van der Velden 2019). Later research that has become a trustworthy source for e-waste statistics estimates that around 62 million metric tons of e-waste arose globally in 2022 (Baldé et al. 2024). Note that if the early figures of hundreds of millions of tons arriving in Agbogbloshie were accurate, then that single site would have to be receiving at least double the amount of e-waste than what the Global E-waste Monitor suggests arose on all of planet Earth in 2022.

Problems, Controversies, and Solutions

One of the reasons sensational numbers like "hundreds of millions of tons" may have been accepted as plausible is how they fit with stereotypical images held by some people in places of dominance over and above people and places who are pushed to the margins. Stereotypes are classification systems that seek to impose order on some piece of social experience and, in doing so, create a distinction between people and places deemed to be "normal" and those deemed to be "deviant." There is a large literature that critically excavates the sedimentation of racist stereotypes about people and places in Africa, Asia, and elsewhere deemed to be peripheral to the core of the so-called West or Global North (Mitchell 2000). There is a long history, for example, of racist representations of "deepest, darkest" Africa as a site of backwardness and primitivism. Representations of e-waste on the continent often trade on such stereotypes (Agyepong 2014; Badcock 2022). Agbogbloshie has repeatedly been described by American and European journalists who characterize the site in biblical terms of hell and damnation. Sometimes media refer to the site as Sodom and Gomorrah, biblical cities destroyed by God and symbols of sin and divine retribution. Tellingly, such representations of Agbogbloshie are offered with little or no reference to the concrete history of the site and how even when such biblical terms are invoked by residents of Accra themselves, it is often so as to make social distinctions between impoverished residents at the site and local elites seeking to assert their economic control of the land (Akese, Beisel, and Chasant 2022).

But there is a flipside to the negative stereotypes associated with e-waste processing. It occurs when people may be more positively portrayed, yet those portrayals retain the two-dimensional flatness of stereotypes. This kind of representation sometimes arises in representations of people and places that process discarded electronics through activities like repair and refurbishment. Their work is sometimes framed as that of a heroic underdog working in a system that, though it has its inequalities, the system's shortcomings can be overcome by individual work effort. Urban studies scholar Ananya Roy (2011) names the problem "slumdog urbanism," a reference to an Oscar-winning film of 2010 that portrays the meteoric rise to stardom of a boy from an impoverished and marginalized section of the Indian city of Mumbai called Dharavi. Roy notes how the story replays myths of self-made individuals who, in the face of seemingly impossible odds, make good through an alchemical combination of luck, pluck, and determination. The power of such a narrative arises from its deflection of criticism of the status quo. The problem, such a story suggests, is not how the system is organized since any of the problems relating to it can be overcome through individual effort within that system. Like the negative side of the stereotypes described earlier, this more positive stereotype similarly forgoes reference to the making of the conditions of people's actual lives and the structural differences of power that uphold some people and places as the dominant norm against which anyone and anywhere that does not match it is formatted as deviant and of lesser value.

The kinds of issues raised by Roy (2011) also appear in criticisms of some framings of the e-waste problem, including when those framings explicitly invoke environmental justice concerns. Critics argue, for example, that not only does the waste dumping story line stand on shaky empirical ground but it also fits into a broader problematic of what some call the white savior industrial complex (WSIC) (see Kherbaoui and Aronson 2021). In short, critics argue that the WSIC brings together initiatives, often well-intentioned, that seek to do "good" for people and places racialized as non-white and typically construed by WSIC organizations and their funders as both exotic "Others" and to be in need of "rescue" that (supposedly) only members of the WISC can provide.

Many initiatives about solving e-waste as an environmental problem also fit into this broader WSIC narrative and its criticisms. It is typical, for example, for American and European ENGO reports and journalism about e-waste to focus on the issue of waste dumping while simultaneously ignoring any possibility that people in poorer parts of the world also have demands of their own for digital technology. The issue here is not about denying that electronic devices imported into countries such as Ghana, India, or Bangladesh never become waste and in need of appropriate management (Burrell 2016). But electronic technologies are not new to countries in Africa or Asia even as places in these regions are all too often still construed as "developing" or "Third World" by people able to inhabit a position of whiteness. The television industry in Nigeria, just to take one example, is more than fifty years old (Esan 2009). No such industry exists without the attendant electronic infrastructure or a sufficient density of consumers with access to electronic technologies like televisions.

Critics of white saviorism's link to the e-waste problem argue that it does harm even as its intentions may be good. Such harms are not abstract. They include criminalizing people who collect and export digital devices discarded in richer markets that with some refurbishment or repair can be sold to satisfy the demand for digital devices in poorer markets overseas. Other harms include providing a kind of reputational cover story for device brand manufacturers who benefit from the emphasis on recycling post-consumer e-waste as a solution. As discussed elsewhere in the book and in relation to Figure 2.1, post-consumer recycling is a necessary but insufficient practice for dealing with pollution and waste arising from the total lifecycle of electronics, including the mining for their materials, and their manufacturing. More robust approaches to e-waste, environmental racism, and environmental justice would, critics argue, start from an orientation where people and places otherwise framed only as victims in need of help from outsiders are, instead, understood to be sovereign agents of their own digital futures. Arguably, what is important in any analysis that seeks to understand how e-waste and racialization operate is an examination of differential power relationships, how they work, and who benefits from differential distributions of specific harms and benefits.

Another important facet of the kinds of issues discussed here is colonialism. Indigenous scholars have evidenced the mutually constitutive relationships between colonialism and pollution. These interventions in how such relationships are conceptualized have important implications for how the pollution and waste arising from the mining for, and the manufacturing of, electronics as well as their use and their afterlives might be mitigated or eliminated. In the words of Max Liboiron (2021), "pollution is colonialism." Such a statement may seem controversial until one considers the foundational linkages between state institutions and companies in countries like Canada or the United States that were founded as settler colonies. Part of the argument this scholarship is making is that many possibilities put forward as solutions to environmental problems take for granted, "settler-colonial jurisdiction over land and life and including Indigenous land and life" (Liboiron and Murphy 2021). When solutions such as regulation of toxic emissions from industrial manufacturing are offered up, they rarely if ever explicitly acknowledge Indigenous sovereignties, even in places where there are extant treaties between Indigenous peoples and settlers. This means that the statement pollution is colonialism is not just theoretical but is an accurate description of situations on the ground, including places such as "Silicon Valley" with deep connections to the electronics industry (Marisa Elena Duarte [Pascua Yaqui] and Jacob Meders [Mechoodpa/Maidu] 2021).

Carbon Fixation

Carbon fixation refers to what some analysts consider to be a problematic over emphasis on CO_2 emissions as the single most important pollution problem or an over emphasis on CO_2 as a sufficient indicator or index of overall waste and pollution problems arising from industry, including electronics manufacturing (TDI Sustainability 2022). To be clear, carbon fixation is not a form of denialism about the climate emergency. Instead, the idea of carbon fixation is meant to capture the importance of a holistic view of pollution and waste arising from industry. For example, as noted earlier, 447 million kg of chemical toxicants were released in the CUSMA region between 2006 and 2020 by electronics manufacturers. Yet, the prevailing practice of electronics companies that publish sustainability reports is to focus on the CO_2 emissions associated with their products. In one sense, companies should be applauded for doing so. However, as the data on chemical toxicant releases suggest, there are many other types of molecules that are released from electronics manufacturing that a singular focus on CO_2 emissions cannot capture. The measurement of CO_2 emissions cannot capture the effects of those other toxicants, nor can it capture the contributions to the climate emergency of several important gases other than CO_2 that are used in the manufacturing of electronics. In these ways, carbon fixation names two linked problems pertaining to e-waste. The first problem is about how upstream pollution and waste from the manufacturing of electronics are accounted for. The second problem relates to treating that upstream pollution and waste in terms of molecules alone with little or no reference to the industrial organization of infrastructure from which those molecules arise. Two examples can help elucidate the problem of carbon fixation: the role of fluorinated greenhouse gases (or F-GHGs) and per- and polyfluoroalkylated substances (PFAS, sometimes called "forever chemicals").

F-GHGs are a class of gases used in the manufacturing of flat panel displays (i.e., "flat screens") for televisions, computer monitors, tablets, and cellphones. They are important for manufacturing stages that etch patterns into display glass and what are known as chemical vapor deposition tool chambers (CVDs). CVDs rely on specialized vessels and gases like F-GHGs to manufacture components that are important to electronics under vacuum conditions. From the global warming perspective, the issue with F-GHGs is that they are thousands of times more potent greenhouse gases compared to CO_2 (United States Environmental Protection Agency 2024). They also are relatively stable and, consequently, tend

Problems, Controversies, and Solutions

to have long atmospheric lifetimes once they are emitted from manufacturing processes.

The potency and atmospheric lifetime of F-GHGs varies depending on the specific gas under consideration. However, of the F-GHGs used in the manufacturing of flat screens, the one with the lowest global warming potential compared to CO_2 is carbon tetrafluoride (CF_4). CF_4 is 7,390 times more potent than CO_2 as a contributor to global heating and is estimated to have an atmospheric residence time of 50,000 years (United States Environmental Protection Agency 2016). The most potent F-GHG is sulfur hexafluoride (SF_6). Although it has a shorter atmospheric lifetime than CF_4 at 3,200 years, SF_6 is 22,800 times more potent than CO_2 as a greenhouse gas. Nitrogen trifluoride (NF_3) is being used by some flat panel display manufacturers as a substitute for SF_6, but NF_3 is itself a potent greenhouse gas. It has a global warming potential 17,200 times higher than CO_2 and an atmospheric lifetime estimated to be 740 years.

Table 2.2 shows the atmospheric lifetime in years and the global warming potential over 100 years of different gases used in two key phases of manufacturing screens ("flat panel displays") for electronic devices. Carbon dioxide (CO_2) is the benchmark against which each of the other gases is compared. Tick marks show in which phase or phases in the manufacturing process that a given gas is used. Note that the shortest atmospheric lifetimes of these gases amount to

Table 2.2 Gas Applications and Climate Impact in Flat Panel Manufacturing

Compound	Chemical Vapor Deposition (CVD) Chamber Cleaning	Plasma Etching	Atmospheric Lifetime (years)	Global Warming Potential (100 years)
CO_2	N/A	N/A	Variable	1
C_2F_6	√	√	10,000	12,200
CF_4		√	50,000	7,390
SF_6	√	√	3,200	22,800
NF_3	√	√	740	17,200
CHF_3		√	270	11,700
C_3F_8	√		2,600	8,830
$c\text{-}C_4F_8$	√	√	3,200	10,300

Source: United States Environmental Protection Agency (2016).

centuries, but some are estimated to remain in the atmosphere for millennia and have global warming potentials many thousands of times higher than CO_2. No amount of pollution and waste mitigation that occurs after consumers repair or recycle their devices can recoup pollutants and waste released before consumers even purchased their devices, that is, in the manufacturing stage.

PFAS are another class of chemicals with important toxic consequences. Exposure to PFAS "has been linked to deadly cancers, impacts to the liver and heart, and immune and developmental damage to infants and children" (United States Environmental Protection Agency 2024). Like F-GHGs, the implications of PFAS are missed if carbon fixation takes hold of industry assessments of the pollution and waste arising from it. PFAS are used to manufacture electronics, but they also occur in the materials making up devices themselves. Common sites of PFAS in electronic devices include printed circuit boards, capacitors, components that convert mechanical, photo, or thermal signals into electrical ones—such as microphones, speakers, and touchscreens—displays, wiring, cables, lithium-ion batteries, and in surface coatings for device housings to reduce scratching and erosion. To manufacture electronics is to release PFAS into the environment and to use an electronic device is to come into contact with PFAS.

Newly available data for PFAS releases from the electronics sector are available from the US Environmental Protection Agency (EPA) (n.d.). These data offer important, if limited, insights into PFAS releases from the electronics manufacturing sector in the United States. The limits to these data are substantial. At the time of writing there are over 7.7 million PFAS compounds registered by the US National Institutes of Health (PubChem 2024). Yet, US-based manufacturers are required to report releases of only 189 individual PFAS compounds, a number that represents merely 0.002 percent of the 7.7 million PFAS known to exist. Nevertheless these newly available data disclose that the electronics sector is responsible for over 88,000,000 pounds (approximately 39,916,128 kg) of PFAS releases per year (United States Environmental Protection Agency 2023, 33).

There are multiple segments of the manufacturing of electronics that rely on PFAS and substitutes exist for only some of them. This is particularly true with respect to the manufacture of semiconductors or what have been called the building blocks of all electronics. PFAS compounds are used to control the photolithography processes, rinsing, etching, and several other phases of semiconductor manufacturing. For most or all of these processes, no alternatives to PFAS appear to exist as yet, although some may be under development (Lay et al., n.d.).

Thinking carefully about the concept of carbon fixation is important as a reminder of the work done by problem framing. As discussed earlier, when e-waste is framed as a problem of what happens after consumers get rid of their devices, that framing excludes or brackets out those other parts of the story that are far more consequential in terms of waste and pollution such as the mining for, and the manufacturing of, electronics. Similarly, carbon fixation is a reminder that framing the problems associated with the environmental impacts of electronics primarily in terms of CO_2 emissions brackets out many other pollution and waste issues associated with electronics.

Solutions: The Waste Hierarchy

Typically, solutions to the e-waste problem are framed in terms of the waste hierarchy. The waste hierarchy is a collection of ideas and strategies for dealing with waste that is probably familiar to readers. The hierarchy presents a list of waste management practices ordered in terms of most to least preferred. The most preferred method is waste prevention. Waste prevention (also sometimes called waste avoidance) is followed by reduction, reuse, recycling, and finally, the least preferred option: disposal. Sometimes additional waste management options are added to the hierarchy. One of the most common is energy recovery. Energy recovery is controversial because it essentially relies on incineration of materials that enter the waste stream, especially plastics, to generate energy. Those processes, however, release CO_2 and other toxicants. While the waste hierarchy is a familiar way of thinking about solutions to waste problems, it is also relatively new. US environmental historian Martin Melosi traces the emergence of the waste hierarchy to broader policy shifts taking place in the United States in the late 1980s (Melosi 2005). The hierarchy did not become formally incorporated into European legislation until 2008 (Torsello 2012).

In principle, the waste hierarchy is a generic approach to waste management that applies to all phases of economic production (i.e., raw material extraction, manufacturing, and use). In practice, however, most of the attention on the waste hierarchy is confined to post-consumer waste management practices. This focus on downstream, post-consumer waste management practices is in part a consequence of limited data collection on waste disposition in mining and manufacturing. It may be surprising to learn, for example, that the US EPA has no mandate to collect data on non-hazardous waste generation and disposal in manufacturing (MacBride 2012). The latter situation is important to keep in mind when analyzing what data are publicly reported about pollution and waste arising from electronics manufacturing. Typically, those data refer only to types of pollution and waste deemed hazardous, rather than non-hazardous. What that means is the many millions of kilograms of chemical toxicants that are reported to be released from electronics manufacturers in North America are only a partial count of all waste arising from those same manufacturers. Data for pollution and waste arising from all the mining needed for electronics are similarly partial. Data can be found for some specific mines, but sector-wide reports of data are harder to come by. Moreover, attributing pollution and waste arising during mining to the electronics manufacturing sector is challenging since raw material extraction of metals has multiple industrial uses even in instances where electronics represent a significant market for this or that material.

In the following sections, each phase of the waste hierarchy is discussed in turn, but in the reverse order of preference moving from disposal through to waste avoidance and prevention. Organizing the sections in this reverse order is helpful because the more preferential aspects of the waste hierarchy—reduction, avoidance, and prevention—lead to broader discussions of how the industrial organization of mining for, and manufacturing of, electronics might be reorganized to substantively mitigate or eliminate pollution and waste arising from the sector.

Disposal

For Baldé et al. (2024) disposal of e-waste is a sub-optimal method of treatment. It results in the loss of resources and can have negative environmental impacts even in well-engineered landfills. Indeed, most e-waste legislation forbids disposal as a treatment method. Nevertheless, Global E-waste Monitor data indicate that of the 62 Mt of e-waste arising globally in 2022, 14 Mt or about 22.5 percent, is estimated to be disposed of in landfills in high- and upper-middle-income countries globally (Baldé et al. 2024, 11, 33, 35). Fifty percent of those 14 Mt are lost from viable recovery. Meanwhile, 18 Mt of e-waste are estimated to be collected outside formal channels in low- and middle-income countries of which 2 Mt (or 10 percent) go to uncontrolled disposal.

Problems, Controversies, and Solutions 75

managing the waste arising from their products and are prohibited from passing those costs on to consumers of their devices would be more likely to achieve the expressed goals of EPR and the polluter-pays-principle.

E-waste recycling is not a panacea, but nor is it a useless distraction from other more effective ways to manage pollution and waste arising from electronics. A more nuanced understanding of e-waste recycling is needed. One issue that is important to understand is that when the word "recycling" is used in the context of e-waste management it is almost always by default referring to post-consumer materials recovery, rather than recycling at any other phase in the lives of electronics, that is, mining or manufacturing. One of the reasons for this is that data on post-consumer recycling is typically much more readily available than for recycling in any of the industrial phases of electronics related to mining or manufacturing. It is not that data about materials reuse in the mining for and manufacturing of electronics do not exist. Companies involved in those phases of industrial production keep detailed records about materials and energy used and reused in their processes. However, those data are rarely released publicly. Moreover, when they are released publicly, for example when companies publish sustainability reports, there is a limited ability to independently verify those data.

To gain a more nuanced understanding of recycling and how it relates to e-waste it is helpful to distinguish recycling into three broad categories: (1) recovering recyclable materials, (2) use of recovered materials by manufacturers, and (3) the purchase and use of products made with recovered materials (M. Smith 1997). Making these distinctions assists in getting beyond unhelpful dichotomies between policy responses devoted to either demand- or supply-side supports for recycling. Supply-side policies tend to assume that once recycled materials become cheap(er), plentiful, and of sufficient quality, manufacturers will automatically substitute them for primary raw materials they would otherwise use. Demand-side policies try to make favorable conditions for recycled products based on the assumption that large enough demand (e.g., spurred via public procurement) will automatically lead to manufacturers satisfying that demand.

Both supply-side and demand-side policy approaches tend to simply assume that the second category of recycling—the use of recovered materials by manufacturers—will automatically result. Unfortunately, for a variety of reasons the use of recovered materials in industry is just not that simple. One reason arises from the specific geographical distribution and location of infrastructure and supply chains. If the materials to be recovered for use by manufacturers are to be derived from post-consumer waste streams, then the location of where most consumers use and dispose of their devices do not necessarily match up with the industrial geography of electronics manufacturing. Where such geographical mismatch occurs, supply chains that would bring materials recovered from post-consumer devices to sites of manufacture have to be linked up. Doing so may mean shipping along multiple transport modes over substantial distances, including export from regions of electronics consumption to regions of electronics production.

Another important consideration is how well-recovered materials do or do not work with specific production processes and technologies. The kinds of industrial machinery involved in electronics manufacturing tend to have very specific and narrow tolerances for variations in the quality of material inputs. Because of the complexity of manufacturing electronics, it is difficult to generalize on this point, but the basic idea is that just because recovered materials are made of a specific metal (e.g., aluminum), that does not automatically mean that the recovered materials match the tolerances for quality that a given manufacturing process requires. Smith's (1997) study of the US paper industry found this to be the case even in an industry where the inputs may not intuitively seem to be subject to the same kinds of technological complexity as electronics. From a materials point of view, electronics are even more complex. They combine plastics, metals, and glasses into a single finished product. Moreover, making electronics involves manufacturing processes and machinery that have extremely narrow tolerances for variations in materials quality. Taken together, what all this means is that the second category of recycling—the use of recovered materials by manufacturers—by the electronics sector can be challenging (if not impossible).

Another facet of the nuance with which recycling needs to be considered is the degree to which it may or may not actually result in the reduction of raw material extraction and/or in the prevention of environmental damage. Waste management and recycling expert Samantha MacBride (2019) describes this issue as the distinction between conservation and preservation. Conservation is what happens if recycling reduces the extraction of raw materials. Preservation is what happens if recycling prevents the degradation of ecosystem complexity. MacBride questions whether typical recycling practices lead to actual conservation or preservation. Like Smith (1997), MacBride (2019) argues that a variety of assumptions have to hold true if genuine conservation and preservation are to be achieved through standard recycling practices. For example, genuine preservation is not achieved if the materials deemed to be a raw material stock,

Problems, Controversies, and Solutions

say, bauxite from which aluminum is extracted, is put into use for a previously nonexistent product, say cellphones, while aluminum scrap is recycled back into aluminum beverage containers. Cellphones only became commercially available in the early 1980s. Aluminum drink cans have been around much longer. Even if all of the aluminum drink cans put on the market are recycled, there will not be enough aluminum from those cans to then be used for both all the cans to be manufactured from recycled aluminum scrap and novel devices like cellphones being put on the market. Another example comes from plastics which are also important materials for electronics. Major fossil fuel companies tout recycling as a solution to plastic pollution. An existing attempt to recycle plastics back into their constituent chemicals has managed to process 500 kt of plastics back into their constituent chemicals (Paben 2022). This amount, 500 kt, may sound substantial, but it is not even 1 percent of annual new plastics production in the United States.

Another important assumption built into the hope that recycling will lead to preservation and/or conservation is that it is technically possible to recycle all materials of which electronics are composed. This is not the case (Yang et al. 2012). Electronics often incorporate composite materials. These are amalgams of various plastics, metals, and glasses that once fused chemically are unable as yet to be disaggregated back into their constituent materials. This may change in the future, but currently this is pure speculation. Electronics recycling as it exists now cannot come close to recovering 100 percent of all of the various materials electronics are made of. What this means is that the recycling of electronics is never complete (this is also the case with other product types). As a consequence, electronics recycling always results in what the industry calls residuals. Those residuals are comprised of a mix of materials that have to be managed somehow. Typically, the options are disposal or incineration. Neither of those options makes those materials simply disappear. Disposal simply sequesters them temporarily in some location while incineration may transform them chemically into nontoxic (though just as often merely less toxic or differently toxic) solids and gases that also need disposal or dispersal into the atmosphere.

Despite hopes to the contrary, recycling cannot fulfill the wish of instituting a genuinely circular economy that conserves resources and preserves ecosystems over the long term. Electronics are too materially complex to achieve 100 percent material recovery. Meanwhile, the mining, the making, and the using of electronics take place within a broader economic context premised on continual growth. An economy organized on the principle of perpetual growth requires new inputs of materials and energy that even perfect recycling cannot deliver (De Decker 2018). Such limits to recycling as a solution to electronic waste, however, do not mean that recycling is a hopeless pursuit. Post-consumer recycling can achieve certain goods such as diversion of materials from landfills and the creation of some employment. Recycling implemented in the mining and manufacturing stages associated with electronics can, in principle, reduce the throughput of materials and energy in the sector. It is crucial to keep in mind, however, that a reduction of material and energy throughput in one sector, such as electronics, is not the same as an overall reduction in those throughputs. Moreover, as the earlier discussion of the Jevons Paradox suggests, reductions in material and energy throughput tend to be short-lived, the gains achieved typically being wiped out by subsequent growth.

Reuse

Direct reuse of devices that users already have is the most environmentally beneficial action that an individual can personally take. This means using devices that a user already has access to for as long as possible. Existing devices that users have access to also embody the materials and energy that went into making those devices in the first place. Again, it is the mining for, and manufacturing of, electronics that generate the largest majority of pollution and waste. Therefore, direct reuse is a way to amortize the energy and materials embodied in devices, including pollution arising from them, for as long as possible. In this sense, direct reuse is a form of material and energy conservation.

There are, of course, limits on what reuse can accomplish, but the common refrain that electronics quickly become obsolete especially as a consequence of changes in software may be somewhat overblown. Especially when it comes to consumer devices like laptops and desktops, there are a number of software options that can extend the useful life of those devices well beyond the typical periods of time they are provided support by the original equipment manufacturer. For example, operating systems such as Ubuntu (a Linux-based operating system) are free and offer support for up to ten years. Moreover, because operating systems like Ubuntu are built to have less demanding system requirements than other common operating systems, they can be loaded and used on older machines thus extending their useful lives. As an example, independent journalist Kris De Decker writing in 2020 was managing to use a second-hand laptop built in 2006

Problems, Controversies, and Solutions

for all of their everyday personal and professional computing needs (Decker 2020). De Decker's example will not be possible for everyone, of course, but it does hint at what is possible in terms of direct reuse.

A key factor that supports direct reuse is repair. Over the last decade or so, substantial advocacy for right to repair legislation in the United States, Europe, and elsewhere has been taking place and has seen some successes. In the United States an organized coalition of trade associations, NGOs, and businesses worked together to form the Repair Association. In 2013, the Repair Association became a nonprofit industry association that lobbies US state and federal governments for right to repair legislation. One tactic their lobbying has been premised on is developing model legislation for states to adopt. The model legislation places certain requirements on original equipment manufacturers (OEMs) to enable do it yourself (DIY) and independent third-party repair of digital devices. These requirements include fair and reasonable terms for purchasing replacement parts, the provision of documents such as circuitry diagrams and other manuals, and access to specialized tools, documentation, and parts that enable the unlocking of devices in order to diagnose problems and repair them (repair.org 2020). A watershed year, 2019 saw twenty US states considering right to repair legislation. Five states (New York, Minnesota, Colorado, California, and Oregon) passed such legislation as of 2024. Moreover, all but two—Wisconsin and New Mexico—are either currently considering such legislation or have done so in the past. Still, lobbying efforts by OEMs against right to repair legislation continue.

The usual counter argument that OEMs use against advocates of right to repair is about security and privacy risks. Cyber security and privacy experts, however, have repeatedly shown that such risks are minimal to nonexistent. The more likely explanation for OEM counter lobbying is that these corporations wish to maintain and even strengthen their monopoly power over all sources of revenue related to their devices, including repair (Doctorow 2021). There is much criticism of this counter lobbying by OEMs. Such critics point out, for example, that in lobbying against right to repair legislation, device manufacturers are undermining a key principle of ownership (Perzanowski 2016). As advocates of right to repair claim, "if you can't repair it, you don't own it" (iFixit n.d.).

One of the ways that OEMs attempt to flout the right to repair is through parts pairing and the use of embedded software. Parts pairing and embedded software essentially create software connections between individual components in a given device. This tactic enables OEMs to invoke the US Digital Millennium Copyright Act that makes anything considered tampering with digital rights management software a federal offense. These tactics enable OEMs to exercise control—and thus extract revenue and profit—over who and what can do repairs of devices under the OEMs' brands. In what may be an important turning point in the fight for right to repair, the latest such legislation passed in 2024 in the state of Oregon. Oregon's regulation outlaws parts pairing but applies only to smartphones while excluding other electronic devices such as video game consoles, medical devices, and cars. Time will tell if the abolition of parts pairing is extended across more device categories if and when more states pass right to repair laws.

Right to repair in the European Union demonstrates both similarities and differences with the approaches taken in the United States. Unlike in the United States, where right to repair legislation is largely a state-by-state affair, right to repair has been folded into EU-wide legislation. In February 2024, new rules were agreed to that bolster right to repair in the EU. These rules cover some similar ground as the model legislation put forward by the Repair Association in the United States, such as reasonable prices for OEM parts, as well as forbidding parts pairing and embedded software that subvert independent repair. The new rules apply to a broader portfolio of devices than is typical under US legislation. The EU rules cover consumer electronics like smartphones and tablets as well as other consumer durables with electronic components such as washing machines, dryers, dishwashers, and fridges.

Other Europe-wide approaches to right to repair include the development of a repairability index for consumer appliances and vouchers for repair. Work toward a repairability index initially took place in France, which mandated such an index be displayed for consumer appliances in the form of a numerical score beginning in 2021. France's efforts became the basis of an EU-wide standardization process begun in 2017 under the auspices of the European Committee for Electrotechnical Standardisation that culminated in Standard EN 45554 "General methods for the assessment of the ability to repair, reuse and upgrade energy-related products" (European Committee for Electrotechnical Standardization 2017). Meanwhile, voucher systems incentivizing consumer repair have proliferated across the region, sometimes at the national scale, sometimes within subnational regions, and in other cases in specific cities (Rezende 2024). While there are differences in the details of each of these voucher systems, their common feature is to provide rebates and discounts for consumers who bring electronic devices in for repair.

Right to repair advocacy is making important gains as the passage of legislation in the United States and the European Union illustrate, but risks

Problems, Controversies, and Solutions 81

to those gains remain. For example, companies may continue to make design decisions that reduce or eliminate repairability itself. If such a situation comes to pass, one may have the legal right to repair a device, but that right is limited or negated if repair information is not freely available, spare parts are unavailable and or unaffordable, tools are inaccessible, and the device itself cannot be disassembled and reassembled without destroying it (iFixit 2024). Non-destructive disassembly and reassembly have become a particularly acute issue as device sizes have shrunk in combination with the use of adhesives rather than fasteners. Apple's wireless earbuds have become symbolic of these broader industry trends (Haskins, Koebler, and Maiberg 2019). AirPods typically lose the ability to hold a charge after approximately eighteen months. Yet, there is no way to safely separate the battery from the plastic housing. Meanwhile, these earbuds cannot be safely sent for disposal because they contain lithium-ion batteries which can catch fire if they are run through a shredder. Other brands and device models have the same or similar problems. What this means is that even in the jurisdictions where people have the right to repair, they do not have a concomitant right to repairability.

There are ways that risks to the right to repair could be ameliorated or eliminated. One way would be to enforce existing anti-monopoly legislation. Wielding the power of public procurement is another route. The largest purchasers of electronic devices, both historically and currently, are governments. Due to their buying power as large purchasers, public bodies have the ability to write into contracts and purchase agreement requirements that OEMs provide repairable devices. This is not a new idea. For example, military procurement contracts during the US Civil War required equipment to have parts that could be exchanged with those made by multiple manufacturers so as to facilitate repair on the battlefield (M. R. Smith 1985).

Reduce, Avoid, Prevent

To reduce, avoid, and prevent waste and pollution arising from electronics, far more attention needs to be devoted to what happens before devices are discarded by their users. That means turning attention upstream toward the mining for and manufacturing of electronics. Consider again Figure 2.1. Strategies to reduce waste and pollution from electronics that focus on post-consumer discards, tacitly or otherwise, are content to deal with that waste and pollution only after they already exist. Moreover, though such downstream strategies are needed, they are not sufficient on their own to deal with overall waste and pollution arising from the electronics sector. They also risk tacitly or otherwise bolstering the power of device manufacturers to pass off the costs and harms of overall pollution and waste from the sector to broader society while the manufacturers enhance their power to accumulate profits. As Figure 2.1 illustrates, finding ways to reduce mining waste arising at a single copper mine by 1 percent would be equivalent improving the post-consumer recycling of discarded electronics in the EU by 28 percent. Meanwhile, reducing the waste arising from electronics manufacturing in the EU would be like improving post-consumer recycling of devices by almost 5 percent in the EU. Advocates of efficiency, then, should favor not just the recuperation of materials and energy from downstream post-consumer recycling but also the reduction, avoidance, and prevention of pollution and waste arising upstream before devices are discarded by their users.

There is a more fundamental point to be made here about how public advocacy and regulation relate to the organization of democratic and market societies. Scholar of waste studies Zsuzsa Gille (2007) asks her readers to consider what good are democratic institutions for asserting rights to a livable environment if the very best they can achieve is to remediate pollution and waste only after they have occurred rather than prevent them in the first place? This is a profound question that extends well beyond the case of e-waste. At the same time, e-waste offers a charismatic case with which to examine these kinds of fundamental issues. There are certainly challenges toward doing so, but just because something may be difficult to do does not mean it should not be done.

One important avenue by which to reduce, avoid, and prevent pollution and waste from electronics is public advocacy for the reduction and elimination of toxicants used in electronics manufacturing processes and embedded in the devices themselves. There is a long history of community and worker activism going back to the earliest days of Silicon Valley geared toward these goals. The Silicon Valley Toxics Coalition (SVTC), for example, formed in the 1970s in milieu of community and worker concerns about links between chemical harms to electronics workers in factories and the contamination of groundwater in the region. As already discussed, chemical toxicants of many kinds are used in the manufacturing of electronics. These toxicants can pose direct risk to workers in the factories as well as to broader communities when those toxicants and/or their by-products are disposed of. It was common practice for companies to dispose of such toxicants through manufacturing plants' connections to municipal sewage systems. Some of these chemicals eventually ate through drainage pipes and leaked into the surrounding subsurface and its ground water. The action of SVTC and other allied organizations eventually led to the phasing out of some—though by no means all—toxicants from the electronics manufacturing processes in the region. Over the course of decades, public advocacy also led to parts of Silicon Valley being designated as Superfund sites. Such designation meant public money being devoted to remediation of contaminated sites. This work is ongoing to this day and, as already noted, will require centuries more to reach what are considered safe levels of exposure to toxicants in drinking water by the standards of today's US EPA. Similar public advocacy emerged wherever electronics manufacturing can be found and continues to this day.

A second key avenue by which to reduce, avoid, and prevent pollution and waste from electronics is public policy. An important example of such public policy is the EU's Reduction of Hazardous Substances (RoHS) legislation. RoHS targets the electronics sector specifically. It came into effect in 2006 and, as its name implies, it aims to reduce or eliminate the use of specific toxicants in electronics manufacturing and in devices themselves. The legislation restricts the use of ten substances including heavy metals like lead, mercury, and cadmium as well as other chemical toxicants that are typically used as flame retardants in device components such as wiring. The significance of RoHS and other legislation like it is more than just that they reduce or eliminate the use of particular toxicants in electronics manufacturing and devices. Also significant is where in the overall lifecycle of electronics this public policy intervention occurs, that is, in manufacturing. This insertion of public decision-making in manufacturing is important because it intervenes in the chemical fate given to devices that consumers will eventually discard. That means that RoHS is intervening to change how that which will become post-consumer waste is made in the first place. In this sense, RoHS is doing more than merely remediating pollution and waste after they already exist. It is also in a measurable sense avoiding and preventing such pollution and waste from arising in the first place. Since the EU adopted RoHS legislation several other countries important to the global production of electronics have adopted similar legislation, notably China, South Korea, and Singapore. China has gone further than other jurisdictions with its RoHS legislation by requiring manufacturers to publicly disclose how their products comply with the legislation. There is evidence to suggest that the mere requirement to disclose such information leads to companies finding ways to reduce or eliminate toxicants from their supply chains and from their products (Ma Jun et al. 2018).

Other examples can be found in China's law called the Inventory of Existing Chemical Substances Produced or Imported in China (IECSC), the EU's Registration, Evaluation, Authorization and Restriction of Chemicals (REACH) legislation, and a recent reform of the United States' Toxic Substances Control Act (TSCA). IECSC was introduced in 2003. Its most recent iteration lists more than 4,700 chemical substances that must be reduced, substituted for with less toxic alternatives, or eliminated. The EU's REACH legislation was introduced in 2006 and is far more ambitious than RoHS. REACH covers all manufactured products imported into the EU, not just electronics. It also seeks to phase out toxicants wherever possible. That means requiring chemical manufacturers to register with the European Chemical Authority and that information about chemicals be available to all users of those chemicals, including downstream manufacturers who use them in their manufacturing processes and/or incorporate them into their final products. In the United States, the reform of TSCA in 2016 gave new powers to the EPA. Prior to this reform, the EPA had no mandate to evaluate chemicals for toxicity. With the reforms the EPA now has the authority not only to assess chemical hazards but also to require their elimination. The broader significance of IECSC, REACH, and TSCA is that they are concrete examples of public decision-making moving upstream to change how products that will eventually be discarded are made in the first place. In other words, these pieces of public decision-making work to avoid and prevent toxic pollution before it occurs or, at the very least, reduce it during the manufacturing of devices and reduce toxicants being incorporated into those devices.

While examples like RoHS, IECS, REACH, and TSCA are promising developments, they are also not panaceas. All of these pieces of legislation make room for the use of some chemicals in specific applications where non or less-toxic alternatives do not exist. The electronics manufacturing sector has many such chemical pinch points for which substitutes have yet to be found. This is particularly true with F-GHGs discussed earlier and with a broad category of chemicals called per- and polyfluoroalkyl substances (PFAS). PFAS are sometimes known as forever chemicals because they do not break down under typical ecological conditions. They are entirely synthetic meaning they only exist as a consequence of technological and scientific interventions. Yet, the toxicology of PFAS remains significantly underdeveloped relative to the many millions of PFAS that already exist and are available for industrial use.

Problems, Controversies, and Solutions

Some PFAS chemicals are known to pose significant toxic risks, but the vast majority of PFAS have not even been tested yet. Authors of a recent report for ChemSec, a nonprofit devoted to finding nontoxic chemicals for industrial applications, were unable to find alternatives for the use of PFAS in at least five key phases of semiconductor manufacturing: rinsing solutions, edging, wafer thinning, vacuum pumps, vapor phase soldering, and protective coatings for semiconductor components (Fong et al. 2023; Lay et al. n.d.). Some alternatives could be found in other parts of the electronics manufacturing process, yet even here several of the alternatives substitute "short-chain" PFAS for "long-chain" ones with unknown capacities for toxicological harm. As toxicologists point out, despite the hope that short-chain PFAS are less- or nontoxic, sufficient toxicological research has yet to be completed to know for sure (Ritscher et al. 2018; Scheringer et al. 2014). So, the prospects for alternatives to PFAS used in the manufacturing of electronics are mixed, and without semiconductors there are no electronics. The concerns about PFAS releases into the environment from semiconductor manufacturing are gaining public attention in the United States and elsewhere, especially as the CHIPS Act provides public money to semiconductor manufacturers to build new facilities in the United States. These facilities will release PFAS and other chemical toxicants with their wastewater discharges into public water treatment infrastructure in the communities where they are located.

The situation with PFAS points to the continued, even renewed, significance of an older form of public policy relevant to pollution and waste arising from the electronics sector: pollution release and transfer registries (PRTRs). PRTRs are publicly available inventories of chemical substances released from industrial manufacturing facilities, including those related to the electronics sector. As with other regulatory requirements for disclosure, some evidence exists that requiring industries to publicly disclose information about their emissions of pollution can lead those industries to finding ways to reduce those emissions (Ma Jun et al. 2018). PRTRs could be modified to expand their utility for reducing and/or eliminating pollution and waste arising from the electronics sector. For example, under international trade agreements such as NAFTA and the EU, companies are already mandated to disclose lists of pollutants emitted from their manufacturing facilities in those regions. At present, however, those lists of pollutants are not harmonized. Indeed, out of the hundreds of chemicals required to be disclosed in these regions, only ten such chemicals are shared between them. Concretely, that means that a manufacturer that operates facilities in NAFTA and in the EU will look very different from a pollution release perspective based on data from each of those regions. Harmonizing the rules for disclosure between jurisdictions would be an important step toward further mitigating the pollution and waste arising from the electronics manufacturing sector.

Another route for improvement could follow the example Section 1502 of the US Dodd–Frank Act by applying its principles to existing PRTR legislation in the United States. Section 1502 requires any company operating within the United States to disclose whether its manufacturing processes and/or products contain what have been called "conflict minerals" originally mined in the Democratic Republic of Congo. What Section 1502 suggests is that it is possible to extend reporting requirements well beyond the boundaries of a nation such as the United States to incorporate the global production networks and supply chains of companies wherever they may operate. It is possible to imagine, then, that PRTR reporting requirements that exist in the United States, for example, could be extended throughout the global production networks of any electronics manufacturer with operations within the United States. Doing so would mean that a company operating in Silicon Valley (or elsewhere in the country) must merely disclose the same list of pollutants throughout its production networks and supply chains wherever they may be located outside the United States. The kinds of public policy reforms discussed here are unlikely to fully avoid and prevent pollution and waste from the electronics sector, but they may help to reduce them.

Design Solutions

How electronic devices are designed plays an important role in their overall service life. Electronics design is a multidisciplinary field that brings together a variety of expertise. Devices are a result of decisions made by designers, all with different priorities. An electronics engineer might be concerned with lengthening battery life. A materials scientist might be looking for ways to make screens more resistant to shattering. An industrial designer might hope to deliver an aesthetically beautiful user experience. All of this means that there may be both complementary and competing goals amongst designers of electronic devices. Finding ways that designers of different expertise can support the goals of reducing and eliminating pollution and waste arising from electronics is challenging, but some broad guidelines can be identified. Several of these guidelines also cut across various layers of the waste hierarchy.

Problems, Controversies, and Solutions 87

First, designers could prioritize device repair and reuse. Doing so would mean avoiding design decisions that make it difficult or impossible to fix devices if and when they break. One common design decision that seems to have prioritized aesthetics over repair and reuse is the design of sealed or fused form factors. In the design world "form factor" refers to the outward physical appearance of a device (e.g., its size, shape, and style). Sealed or fused form factors are familiar across many classes of electronic devices including phones, laptops, monitors, and tablets. Such form factors are typically achieved using adhesives rather than fasteners (e.g., bolts, screws) to hold devices together as a sealed unit. Not only may the adhesives themselves be toxic, but their use means that opening devices for repairs may be impossible or even destroy the device in the process. In this sense, sealed form factors tend to shorten the useful lives of devices and thus forgo opportunities to conserve the materials and energy embodied in those devices.

Designers might find ways to reduce the mix of materials used in devices. Greater material heterogeneity can increase the challenge of recycling, especially when materials form amalgams for which there is, at present, no technical way to later separate them. Reducing material heterogeneity also reduces the risk of cross-contamination of materials that are recovered via recycling. For example, different plastics used in the same device can be difficult if not impossible to separate from one another. They may also reduce the quality of non-plastic materials recovered through the recycling process by becoming co-mingled through shredding and incomplete sorting. Designers can work together to examine what combination of materials and assemblies enhances the durability, repairability, and remanufacturing options of devices. This is a challenging but also highly creative process in which collaboration across design fields can potentially lead to innovations in device manufacturing processes and design.

Designers might also look to how they can make more modular devices. Modularity refers to the degree to which devices have sub-units and/or components that can be swapped out for others. More modular devices mean that parts and components can be replaced if and when a user breaks them or wishes to upgrade, without having to replace the device as a whole. For example, when designers decide to solder memory chips to motherboards, they have also decided that those same memory chips cannot be replaced or upgraded. In contrast, designers that provide for slotted memory (i.e., random access memory chips that can be pulled out and replaced by hand) have made a design decision that makes their device more modular and thus potentially longer lasting. Longer lasting devices conserve the energy and materials embodied in them.

One common design decision that should be avoided is serialization (also called parts pairing). Serialization uses hardware and/or software identifiers to match device parts. Some brand manufacturers use serialization to lock-in device users and/or repair technicians forcing them to use the same brand manufacturer's parts. This practice is a way for brand manufacturers to enhance their profits, but at the expense of repair and reuse by people who buy their devices as well as independent repair technicians (Purdy 2020). The more difficult repair is made to be the less likely the use life of a device will be extended. Shorter use lives are missed opportunities to conserve the embodied energy and materials and devices over longer periods of time.

Designers can look for ways to use recycled and scrap materials in manufacturing, not just making devices that can be later recycled. The use of recycled materials in manufacturing often offers substantial energy and material savings compared to the use of primary raw materials (i.e., mining). Additionally, materials scientists can search for clean(er)/green(er) product chemistry. Clean(er), green(er) chemistry reduces the risks of toxic harm to people manufacturing devices, device users, people involved in waste management, and environments.

Device designers could shift their overall approach to design problems away from a risk orientation and foster what biologist Mary O'Brien (1993) calls alternatives analysis. Risk analysis tends to be premised on questions like, "How can health risks from toxic chemical X be reduced by Y percent?" Alternative analysis would ask a different question: "What options are there to replace chemical X with a less- or nontoxic substitute?"

Solutions at the design level can only achieve so much. Designers work in industrial systems. To the extent that those systems are premised on continual growth, solutions offered by designers can be wiped out by that growth. This is once again the rebound effect or the Jevons Paradox. If, for example, a hardware engineer finds a way to use 10 percent fewer PFAS per unit of semiconductors, but 10 percent more semiconductors are manufactured, then the net reduction of PFAS is zero. If the Jevons Paradox comes into play, then alternatives analysis needs to go further and find ways of reorganizing industrial systems. Needless to say, reorganization of an industrial sector is a very challenging problem, but some hints for making progress do exist just as others have yet to be imagined.

Reorganizing for Solutions

The solutions described in the previous sections have limits. One important limit is the degree to which any of them alone or in conjunction could lead to a reduction in the aggregate throughput of energy and materials in the industrial systems that manufacture electronics and give rise to the pollution and waste attributable to the sector. The solutions outlined thus far may go some distance toward reducing energy and material throughput and, in some instances, reducing or eliminating pollution and waste from the sector. Until there are ways to overcome the Jevons Paradox or rebound effect in the sector, a long-lasting reduction to total throughput of materials and energy is unlikely to occur. The solutions proposed so far may slow the rate of pollution and waste arising from the sector. They may eliminate some specific forms of pollution and waste (e.g., when nontoxic chemicals are substituted for toxic ones). However, reducing the rate of growth of material and energy throughput by the sector is at best slow growth, not no growth. Under those conditions rates of pollution and waste arising from the sector may also slow down but they will continue to accumulate nevertheless.

Possible solutions that may lead to more fundamental changes in the pollution and waste attributable to the electronics sector can be glimpsed. One way such change arises from thinking about the difference between efficiency and sufficiency. Efficiency on its own is not enough if the gains it makes are wiped out by growth in the use of devices. Making, distributing, using (and reusing) devices with sufficiency in mind could help. Sufficiency is about the adequacy of something, especially something essential. Not everyone needs to agree that digital devices are, indeed, essential. Yet, it is hard to argue against the claim that these devices are increasingly becoming necessary to perform more and more daily needs by more and more people. Readers could experiment with this idea themselves by seeing how long they can get by without their devices, such as a phone. (Take notes while you try this. Are the situations that arise merely annoying, or do certain daily routines become impossible? The answers are likely to differ for differently situated individuals and groups.)

Obviously, sufficiency is a different orientation to device manufacturing, distribution, and use than is efficiency. Whereas an electronics engineer concerned about efficiency might seek to reduce the amount of electricity needed per central processing unit compute time, the same engineer working from a sufficiency perspective might ask how much electricity is enough to perform this or that task on a device? Someone contemplating the purchase of a new device might ask themselves how much computing power do they need for their typical tasks? They might then make a decision to purchase a used device, rather than the latest model.

Sufficiency need not mean going without. Indeed, a key problem of the contemporary electronics manufacturing sector is that it produces more supply than there is demand. The situation, although not unique to electronics, is so significant that manufacturers and retailers are incentivized to destroy perfectly reusable and even brand new devices (Roberts et al. 2023). Not surprisingly, this is a situation that manufacturers and retailers are reluctant to disclose data on because doing so may harm their brand reputations. However, the European Environmental Bureau (an NGO) has collected media reports and some academic literature describing the destruction of unsold goods, particularly from online retailers. Their estimates suggest that between 600 million and 3.1 billion Euros (>US$647 million to $3.4 billion) worth of electronics purchased online but returned were destroyed in 2021 (Rödig et al. 2021). This study is limited to only returned goods bought and sold on online retail platforms in the EU and excludes items sold in physical retail stores and unsold stock, so while the dollar figures are substantial, they are also underestimates. These dollar figures suggest that there is a substantial surplus of newly manufactured electronic devices that get destroyed every year without ever going into use.

There are four broad reasons that retail returns and unsold goods are destroyed rather than re-distributed for use. These reasons include the underlying business model of manufacturers and retailers as well as legal and policy structures. For some electronic devices, the cost of manufacturing is below the cost of storage and handling of unsold or returned items. Not surprisingly, profit-seeking businesses seek to eliminate costs while maximizing profit. Consequently, where costs of storage and handling exceed the expected return from selling devices, businesses destroy them. Electronics brands and retailers also use destruction to "keep prices high and retain exclusivity" (Roberts et al. 2023, 302). In this sense, the pursuit of profit deliberately produces scarcity where abundance would otherwise exist. Destruction of unsold stock can also result from tax structures and/or legal restrictions and liability. For example, in the United States, under a law first passed in 1789, businesses can claim back 99 percent of the value of imported goods if they are destroyed under the supervision of US customs authorities (Elia 2020; United States Customs and Border Protection 2013, 2). Companies may also fear being exposed to liability risk from returned or unsold devices that, for example, may move and be stored outside the direct control of manufacturers and/or retailers. Laws and policies, however, are rules created by people. As such, they can be rethought and recrafted to achieve more desirable outcomes than incentivizing the destruction of returned and unsold stocks of electronics devices.

A variety of policy tools could be used to reorient the industrial system for making electronics away from product destruction and toward redistribution and (re)use. Roberts and colleagues. (2023) usefully divide the many possible interventions into two broad categories of "upstream" and "downstream" policy approaches. Upstream approaches would be directed toward manufacturers and retailers whereas downstream approaches would be directed toward actors who use and reuse electronics such as consumers and third-party repair businesses. On the upstream side manufacturers and retailers could be offered sales tax rebates when returned and unsold devices are donated to charity or recovered for reuse and redistribution. Manufacturers' devices might be labeled with durability and/or repairability scores; minimum durability, repairability, and modularity may be required of devices and/or longer-term warranties be required upstream and downstream. Retailers could be subject to similar policies as well as levies on unsold and returned volumes of stock that incentivize recovery and re-distribution for reuse rather than destruction (i.e., making it more expensive to destroy than to collect and re-distribute for reuse). Retailers could also be required to publicly disclose data on volumes of unsold goods and how those devices are handled. For consumers there could be the reduction or elimination of sales taxes on buying second-hand or repaired devices as well as various forms of education and awareness campaigns related to the environmental impacts of returns or the benefits of reuse. Third-party repair businesses could have tax reductions or exemptions related to providing services to consumers; they could also be made eligible for rebates on stocking volumes of parts for repair.

These sorts of interventions are often directed toward private market actors, but they need not be inherently constrained as such. For example, there is a long tradition of the postal system in various countries acting as a public service hub for daily needs beyond handling mail such as banking. Indeed, existing public postal systems, when looked at from a different angle, represent massively distributed forward- and reverse-logistics systems that collect and re-distribute a huge variety of products, including electronics. Post offices could be reimagined as community-based collection points for unsold and returned stock that are provided with standardized checks for usability, data security, and the like and then made available to people who want them.

Other reorganizations of industrial systems for making and using electronics are possible, too. Upstream, on the manufacturing side, there are tantalizing possibilities in what sometimes gets called "design global, manufacture local" or DGML (Kostakis et al. 2015). The basic premise of DGML involves ideas or plans for devices that circulate globally via digital means, but where the actual manufacturing takes place at or very near the site of consumption (even, possibly, within peoples' homes). Such local, even "desktop," manufacturing is already possible for some kinds of devices through 3-D printing. The ability to print electronics from semiconductors all the way up to fully functional devices is, at this point, still only possible in highly specialized academic and industrial research laboratories (Thryft 2021). Whether such localized manufacturing could be widely deployed as a substitute for traditional global production networks of electronics is still highly speculative. Nevertheless, DGML represents some promising ways to reorganize the production of electronics toward systems that may be able to minimize or even avoid the kinds of pollution and waste arising from the way electronics are currently manufactured.

Other nascent downstream models can be glimpsed in emerging networks of neighborhood-oriented repair cafés as well as various types of library models. Traditional libraries that lend books are often at the forefront of lending out other categories of devices that people need and want but that otherwise might be too expensive for them to access. Electronics, such as tablets and laptops, are among those categories of devices that are increasingly found in traditional libraries for lending. Other library-like initiatives include those for devices and the tools needed for their repair. These take many forms—everything from whole buildings to secure, distributed networks of lockers organized via an app and supported by minimal user fees (see, for example, LOOPIS n.d., an initiative underway in Sweden). The commonality among the neighborhood-based repair and library models is that they facilitate the repair and/or redistribution and reuse of electronics that already exist. As such, they represent infrastructure for the conservation of materials and energy embodied in those existing devices. This kind of infrastructure is already being used to collect and redistribute devices that, for whatever reason, people might wish to move along and make available to others. Such infrastructure could also be used to collect and redistribute product returns and unsold stock. There is no guarantee that this kind of infrastructure would inevitably lead to aggregate reductions in pollution and waste from electronics, but it does represent possibilities that could be bolstered and extended in ways that do.

Conclusion

Pollution and waste arise at all phases in the lives of electronics, from mining the needed materials, to manufacturing devices, distributing them, using them, and discarding them. Of these various phases it is mining and manufacturing that are most consequential. Yet, much of the public discourse about e-waste frame the issue as being exclusively about what happens to devices after consumers get rid of them. Many of the ongoing problems and controversies associated with e-waste at least partly arise from the way the problem is framed. How such a problem is framed plays an important role in what solutions are considered obvious and what solutions are not considered viable all, while also making some alternatives difficult or impossible to even imagine. It is important, then, for analysts interested in e-waste to think carefully about how pollution and waste arising from electronics is framed as a problem to be solved.

When articulating solutions to pollution and waste problems associated with electronics, it is critical for proponents to think carefully about how well or badly their proposed solution or solutions align with the very different types and proportions of pollution and waste arising in different phases of the lives of electronics. Proponents also need to think carefully about how different actors are placed in relation to relative power to affect change. For example, focusing on implementing a post-consumer recycling program is not going to change the chemistry of the devices that might be recycled in that program. In that sense, there would be a scalar mismatch between the solution proposed and the problem of releases of chemical toxicants from the electronics sector. Just because such a mismatch exists is not an argument against recycling. Instead, the issue is that the solution and the problem are misaligned.

Current trends indicate that pollution and waste arising from all phases of electronics will continue to grow, even if unevenly so. Imaginative interventions that get beyond the status quo of downstream solutions such as post-consumer recycling and right to repair are required. Both are urgently needed, but insufficient on their own to meaningfully mitigate, let alone eliminate, overall pollution and waste arising from the electronics sector.

References

Agyepong, Heather. 2014. "'The Gaze on Agbogbloshie,' Misrepresentation at Ghana's E-Waste Dumpsite." Okayafrica, October 2, 2014. https://www.okayafrica.com/the-gaze-on-agbogbloshie-ghana-heather-agyepong/.

Akese, Grace, Uli Beisel, and Muntaka Chasant. 2022. "Agbogbloshie: A Year after the Violent Demolition." *African Arguments* (blog), July 21, 2022. https://africanarguments.org/2022/07/agbogbloshie-a-year-after-the-violent-demolition/.

Badcock, Jacob. 2022. "Photography after Discard Studies: The Case of Agbogbloshie." *Burlington Contemporary Journal*, no. 7 (November): 3–27. https://doi.org/10.31452/bcj7.discard.badcock.

Baldé, Cornelis Peter, Ruediger Kuehr, Tales Yamamoto, Rosie McDonald, Elena D'Angelo, Shahana Althaf, Garam Bel, et al. 2024. "The Global E-Waste Monitor 2024." Geneva/Bonn: International Telecommunication Union (ITU) and United Nations Institute for Training and Research (UNITAR). https://ewastemonitor.info/wp-content/uploads/2024/03/GEM_2024_18-03_web_page_per_page_web.pdf.

Basel Action Network. 2002. "Exporting Harm: The High-Tech Trashing of Asia." http://www.ban.org/main/library.html.

Basel Secretariat. 2011. "Basel Convention." http://www.basel.int/Portals/4/Basel%20Convention/docs/text/BaselConventionText-e.pdf.

Basel Secretariat. 2021. "Proposal by Ghana and Switzerland to Amend Annexes II, VIII and IX to the Basel Convention on the Control of Transboundary Movements of Hazardous Wastes and Their Disposal." https://resource-recycling.com/resourcerecycling/wp-content/uploads/2021/02/UNEP-CHW-COMM-COP.15-Amendement-AnnexII-VIII-IX-20210119.English-1.pdf.

Basel Secretariat. 2022. "Basel Convention E-Waste Amendments." https://www.basel.int/Implementation/Ewaste/EwasteAmendments/Overview/tabid/9266/Default.aspx.

Basel Secretariat. 2023a. "Development of Technical Guidelines on E-Waste." https://www.basel.int/Implementation/Ewaste/TechnicalGuidelines/DevelopmentofTGs/tabid/2377/Default.aspx.

Basel Secretariat. 2023b. "Technical Guidelines: Technical Guidelines on Transboundary Movements of Electrical and Electronic Waste and Used Electrical and Electronic Equipment, in Particular Regarding the Distinction between Waste and Non-Waste under the Basel Convention." UNEP/CHW/EWG/EWAST.4/2. https://www.basel.int/Implementation/Ewaste/TechnicalGuidelines/Meetings/4theWasteEWGOnlineMar2023/tabid/9490/Default.aspx.

Beaucham, Catherine C., Diana Ceballos, Elena H. Page, Charles Mueller, Antonia M. Calafat, Andreas Sjodin, Maria Ospina, Mark La Guardia, and Eric Glassford. 2018. "Evaluation of Exposure to Metals, Flame Retardants, and Nanomaterials at an Electronics Recycling Company." HHE Report No. 2015-0050-3308. U.S. Department of Health and Human Services, Public Health Service, Centers for

Disease Control and Prevention, National Institute for Occupational Safety and Health. https://doi.org/10.26616/NIOSHHHE201500503308.

Benjamin, Ruha. 2019. *Race after Technology: Abolitionist Tools for the New Jim Code.* Cambridge: Polity.

Burrell, Jenna. 2016. "What Environmentalists Get Wrong about E-Waste in West Africa." The Berkeley Blog, September 1, 2016. https://blogs.berkeley.edu/2016/09/01/what-environmentalists-get-wrong-about-e-waste-in-west-africa/.

Calma, Justine. 2019. "Aluminum Is Recycling's New Best Friend, but It's Complicated." The Verge, September 12, 2019. https://www.theverge.com/2019/9/12/20862775/aluminum-recycling-water-tech-plastic-manufacturing-cocacola-pepsi-apple.

Ceballos, Diana Maria, and Zhao Dong. 2016. "The Formal Electronic Recycling Industry: Challenges and Opportunities in Occupational and Environmental Health Research." *Environment International* 95: 157–66. https://doi.org/10.1016/j.envint.2016.07.010.

Chachra, Deb. 2023. *How Infrastructure Works: Inside the Systems That Shape Our World.* New York: Riverhead Books.

De Decker, Kris. 2018. "How Circular Is the Circular Economy?—Uneven Earth." November 27, 2018. http://unevenearth.org/2018/11/how-circular-is-the-circular-economy/.

Decker, Kris De. 2020. "How and Why I Stopped Buying New Laptops." LOW←TECH MAGAZINE, December 20, 2020. https://solar.lowtechmagazine.com/2020/12/how-and-why-i-stopped-buying-new-laptops/.

Doctorow, Cory. 2021. "Apple's Right-to-Repair U-Turn." *Medium* (blog), November 21, 2021. https://doctorow.medium.com/apples-right-to-repair-u-turn-e678cf138f74.

Elia, Ariele. 2020. "Fashion's Destruction of Unsold Goods: Responsible Solutions for an Environmentally Conscious Future." *Fordham Intellectual Property, Media and Entertainment Law Journal* 30, no. 2: 539.

Elliott, Bobby. 2015. "Kuusakoski to Replace ADC Option with Storage Cell." *E-Scrap News* (blog), March 12, 2015. https://resource-recycling.com/e-scrap/2015/03/12/kuusakoski-to-replace-adc-option-with-storage-cell/.

Esan, Oluyinka. 2009. *Nigerian Television: Fifty Years of Television in Africa.* Princeton, NJ: AMV Publishing Services.

European Committee for Electrotechnical Standardization (CENELEC). 2017. "CEN-CENELEC TC10 Material Efficiency Aspects for Ecodesign." Secretary Enquiry (New Work Item 65685 / prEN 45554), General Methods for the Assessment of the Ability to Repair, Reuse and Upgrade Energy Related Products. [DRAFT]. https://www.eera-recyclers.com/files/cen-clc-tc10sec135dc-sec-enq-pren45554-repare-reuse-upgrade-2.pdf.

Eurostat. 2019a. "Manufacture of Computer, Electronic and Optical Products, Electrical Equipment, Motor Vehicles and Other Transport Equipment." https://ec.europa.eu/eurostat/data/database.

Eurostat. 2019b. "Waste Electrical and Electronic Equipment (WEEE) by Waste Management Operations | Waste Collected from Households." https://ec.europa.eu/eurostat/data/database.

Fong, Art, Monika Roy, Catherine Rudisill, and Joel Tickner. 2023. "Using Alternatives Assessment to Support Informed Substitution of PFAS in the Electronics Industry." Association for the Advancement of Alternatives Assessment (A4). https://saferalternatives.org/assets/documents/A4-PFAS-Electronics-August-2023.pdf.

Forti, Vanessa, Cornelis Peter Baldé, Ruediger Kuehr, and Garam Bel. 2020. "The Global E-Waste Monitor 2020: Quantities, Flows, and the Circular Economy Potential." Bonn; Geneva; Rotterdam: United Nations University. https://www.itu.int/en/ITU-D/Environment/Documents/Toolbox/GEM_2020_def.pdf.

Gille, Zsuzsa. 2007. *From the Cult of Waste to the Trash Heap of History: The Politics of Waste in Socialist and Postsocialist Hungary.* Bloomington, IN: Indiana University Press.

Greenpeace International. 2008. "Poisoning the Poor: Electronic Waste in Ghana." greenpeace.org.

Haskins, Caroline, Jason Koebler, and Emanuel Maiberg. 2019. "AirPods Are a Tragedy." *Vice* (blog), May 6, 2019. https://www.vice.com/en_ca/article/neaz3d/airpods-are-a-tragedy.

Huisman, Jaco, I. Botezatu, L. Herreras, M. Kiddane, J. Hintsa, V. Luda di Cortemiglia, P. Leroy, et al. 2015. "Countering WEEE Illegal Trade (CWIT) Summary Report, Market Assessment, Legal Analysis, Crime Analysis and Recommendations Roadmap." Lyon, France.

iFixit. 2024. "Gold Standard." February 22, 2024. https://www.ifixit.com/repairability/gold-standard.

iFixit. n.d. "Self-Repair Manifesto." Accessed January 20, 2021. https://www.ifixit.com/Manifesto.

International Energy Agency. 2022. "A 10-Point Plan to Cut Oil Use." https://www.iea.org/reports/a-10-point-plan-to-cut-oil-use.

Isenberg, Nancy. 2016. *White Trash: The 400-Year Untold History of Class in America.* New York: Penguin.

Jackson, Tim, Jason Hickel, and Giorgos Kallis. 2024. "Confronting the Dilemma of Growth: A Response to Warlenius (2023)." *Ecological Economics* 220 (June): 108089. https://doi.org/10.1016/j.ecolecon.2023.108089.

Jacobs, Mr. Advocate General. 1997. Opinion of Mr Advocate General Jacobs Delivered on 24 October 1996. Criminal Proceedings against Euro Tombesi and Adino Tombesi (C-304/94), Roberto Santella (C-330/94), Giovanni Muzi and others (C-342/94) and Anselmo Savini (C-224/95). European Court.

Kasulaitis, Barbara V., Callie W. Babbitt, and Andrew K. Krock. 2019. "Dematerialization and the Circular Economy: Comparing Strategies to Reduce Material Impacts of the Consumer Electronic Product Ecosystem: Dematerialization and the Circular Economy." *Journal of Industrial Ecology* 23, no. 1 (May 2018): 119–32. https://doi.org/10.1111/jiec.12756.

Keeling, Arn. 2012. "Mineral Waste." In *SAGE Encyclopedia of Consumption and Waste*, edited by Carl Zimring and William L. Rathje, vol. 1, 553–56. Thousand Oaks, CA: SAGE Publications.

Kherbaoui, Jamie, and Brittany Aronson. 2021. "Bleeding through the Band-Aid: The White Saviour Industrial Complex." In *Routledge Handbook of Critical Studies in Whiteness*, edited by Shona Hunter and Christi van der Westhuizen, 269–79. Milton: Taylor & Francis Group. http://ebookcentral.proquest.com/lib/mun/detail. action?docID=6809860.

Klein, Peter. 2009. "FRONTLINE/World Ghana: Digital Dumping Ground." PBS, June 23, 2009. http://www.pbs.org/frontlineworld/stories/ghana804/resources/ewaste. html.

Kostakis, Vasilis, Vasilis Niaros, George Dafermos, and Michel Bauwens. 2015. "Design Global, Manufacture Local: Exploring the Contours of an Emerging Productive Model." *Futures* 73 (October): 126–35. https://doi.org/10.1016/j. futures.2015.09.001.

Lay, Dean, Ian Keyte, Scott Tiscione, and Julius Kreissig. n.d. "Check Your Tech: A Guide to PFAS in Electronics." *ChemSec*. Accessed October 4, 2023. https://chemsec. org/app/uploads/2023/04/Check-your-Tech_230420.pdf.

Leif, Dan. 2019. "Kuusakoski Loses Landfill Option for Leaded CRT Glass." *E-Scrap News* (blog), June 27, 2019. https://resource-recycling.com/e-scrap/2019/06/27/ kuusakoski-loses-landfill-option-for-leaded-crt-glass/.

Lepawsky, Josh. 2020. "Sources and Streams of Electronic Waste." *One Earth* 3, no. 1: 13–16. https://doi.org/10.1016/j.oneear.2020.07.001.

Liboiron, Max. 2014. "Solutions to Waste and the Problem of Scalar Mismatches." *Discard Studies* (blog), February 10, 2014. http://discardstudies.com/2014/02/10/ solutions-to-waste-and-the-problem-of-scalar-mismatches/.

Liboiron, Max. 2015. "How the Ocean Cleanup Array Fundamentally Misunderstands Marine Plastics and Causes Harm." Discard Studies, June 5, 2015. https:// discardstudies.com/2015/06/05/how-the-ocean-clean-up-array-fundamentally- misunderstands-marine-plastics-and-causes-harm/.

Liboiron, Max. 2021. *Pollution Is Colonialism*. Durham, NC: Duke University Press.

Liboiron, Max, and Michelle Murphy. 2021. "Why Pollution Is as Much about Colonialism as Chemicals—Don't Call Me Resilient EP 11." Interview by Vinita Srivastava. http://theconversation.com/why-pollution-is-as-much-about- colonialism-as-chemicals-dont-call-me-resilient-ep-11-170696.

Liboiron, Max, Rui Liu, Elise Earles, and Imari Walker-Franklin. 2023. "Models of Justice Evoked in Published Scientific Studies of Plastic Pollution." *FACETS* 8 (January): 1–34. https://doi.org/10.1139/facets-2022-0108.

Linnell, Jason. 2018. "Heavy Impacts from Lighter Devices." *E-Scrap News* (blog), December 14, 2018. https://resource-recycling.com/e-scrap/2018/12/14/heavy-impacts-from-lighter-devices/.

LOOPIS. n.d. "LOOPIS—ideell förening." Accessed May 7, 2024. https://loopis.org.

Ma Jun, Kate Logan, Ding Shanshan, Ruan Qingyuan, Yuan Yuan, Guo Meicen, and Xu Xin. 2018. "PRTR—Establishing a Pollutant Release and Transfer Register in China: Managing Hazardous Chemicals in Electronics Production." Institute of Public & Environmental Affairs (IPE). https://ipen.org/sites/default/files/documents/IPE%20IPEN%20PRTR%20report_EN_5.9.pdf.

MacBride, Samantha. 2012. *Recycling Reconsidered: The Present Failure and Future Promise of Environmental Action in the United States.* Cambridge, MA: MIT Press.

MacBride, Samantha. 2019. "Does Recycling Actually Conserve or Preserve Things?" *Discard Studies* (blog), February 11, 2019. https://discardstudies.com/2019/02/11/12755/.

Marisa Elena Duarte (Pascua Yaqui), and Jacob Meders (Mechoodpa/Maidu). 2021. "Silicon Valley Is Built on Indian Land." Living Room Light Exchange. https://www.livingroomlightexchange.com/publication.

Masanet, Eric, Arman Shehabi, Nuoa Lei, Sarah Smith, and Jonathan Koomey. 2020. "Recalibrating Global Data Center Energy-Use Estimates." *Science* 367, no. 6481: 984–86. https://doi.org/10.1126/science.aba3758.

Melosi, Martin V. 2005. *Garbage in the Cities: Refuse, Reform, and the Environment.* Pittsburgh, PA: University of Pittsburgh Press.

Mitchell, Don. 2000. *Cultural Geography: A Critical Introduction.* London: Blackwell.

O'Brien, Mary H. 1993. "Being a Scientist Means Taking Sides." *BioScience* 43, no. 10: 706–8. https://doi.org/10.2307/1312342.

Paben, Jared. 2022. "ExxonMobil Explains Its Plans to Scale up Chemical Recycling." *Plastics Recycling Update* (blog), October 12, 2022. https://resource-recycling.com/plastics/2022/10/11/exxonmobil-explains-its-plans-to-scale-up-chemical-recycling/.

Pasek, Anne. 2023. "Getting into Fights with Data Centres: Or a Modest Proposal for Reframing the Climate Politics of ICT." Experimental Media and Methods Lab, Trent University. https://emmlab.info/Resources_page/Data%20Center%20Fights_digital.pdf.

Perzanowski, Aaron. 2016. *The End of Ownership: Personal Property in the Digital Economy.* Information Society Series. Cambridge, MA: MIT Press.

Piatek, Nadine M., Robert R. Seal II, Jane M. Hammarstrom, Richard G. Kiah, Jeffrey R. Deacon, Monique Adams, Michael W. Anthony, Paul H. Briggs, and John C. Jackson. 2007. "Geochemical Characterization of Mine Waste, Mine Drainage, and Stream Sediments at the Pike Hill Copper Mine Superfund Site, Orange County, Vermont." Scientific Investigations Report. Reston, VA: United States Geological Survey. https://pubs.usgs.gov/sir/2006/5303/.

PubChem. 2024. "PubChem: PFAS and Fluorinated Compounds in PubChem: PFAS Breakdowns by Chemistry." https://pubchem.ncbi.nlm.nih.gov/.

Purdy, Kevin. 2020. "Is This the End of the Repairable iPhone?" *iFixit* (blog), October 29, 2020. https://www.ifixit.com/News/45921/is-this-the-end-of-the-repairable-iphone.

Reno, Joshua O., and Britt Halvorson. 2022. "Waste and Whiteness." In *The Routledge Handbook of Waste Studies*, edited by Zsuzsa Gille and Josh Lepawsky, 41–54. London: Routledge.

repair.org. 2020. *Model State Right-to-Repair Law.* https://static1.squarespace.com/static/53821f30e4b07bcdae103594/t/6007268d2e8b8444b6b064e3/1611081357764/2021+Model+R2R+Bill+%283%29.docx.

Rezende, João Antonucci. 2024. "A Comprehensive Overview of the Current Repair Incentive Systems: Repair Funds and Vouchers." *Right to Repair Europe* (blog), March 11, 2024. https://repair.eu/news/a-comprehensive-overview-of-the-current-repair-incentive-systems-repair-funds-and-vouchers/.

Ritscher, Amélie, Zhanyun Wang, Martin Scheringer, Justin M. Boucher, Lutz Ahrens, Urs Berger, Sylvain Bintein, et al. 2018. "Zürich Statement on Future Actions on Per- and Polyfluoroalkyl Substances (PFASs)." *Environmental Health Perspectives* 126, no. 8: 084502. https://doi.org/10.1289/EHP4158.

Roberts, Hedda, Leonidas Milios, Oksana Mont, and Carl Dalhammar. 2023. "Product Destruction: Exploring Unsustainable Production-Consumption Systems and Appropriate Policy Responses." *Sustainable Production and Consumption* 35 (January): 300–12. https://doi.org/10.1016/j.spc.2022.11.009.

Rödig, Lisa, Dirk Jepsen, Till Zimmermann, Robin Memelink, and Anna Falkenstein. 2021. "Policy Brief on Prohibiting the Destruction of Unsold Goods." Brussels: European Environmental Bureau, Institut für Ökologie und Politik GmbH.

Roy, Ananya. 2011. "Slumdog Cities: Rethinking Subaltern Urbanism." *International Journal of Urban and Regional Research* 35, no. 2: 223–38. https://doi.org/10.1111/j.1468-2427.2011.01051.x.

Rust, Susanne, and Matt Drange. 2014. "Cleanup of Silicon Valley Superfund Site Takes Environmental Toll." The Center for Investigative Reporting, March 17, 2014. https://www.revealnews.org/article/cleanup-of-silicon-valley-superfund-site-takes-environmental-toll-2/.

Sander, Knut, Stephanie Schilling, Naoko Tojo, Chris van Rossem, Jan Vernon, and Carolyn George. 2007. "The Producer Responsibility Principle of the WEEE Directive." ec.europa.eu/environment/waste/weee/pdf/final_rep_okopol.pdf.

Scheringer, Martin, Xenia Trier, Ian T. Cousins, Pim De Voogt, Tony Fletcher, Zhanyun Wang, and Thomas F. Webster. 2014. "Helsingør Statement on Poly- and Perfluorinated Alkyl Substances (PFASs)." *Chemosphere* 114 (November): 337–39. https://doi.org/10.1016/j.chemosphere.2014.05.044.

Smith, Maureen. 1997. *The U.S. Paper Industry and Sustainable Production: An Argument for Restructuring.* Urban and Industrial Environments. Cambridge, MA: MIT Press.

Smith, Merritt Roe. 1985. "Army Ordnance and the 'American System' of Manufacturing, 1815–1861." In *Military Enterprise and Technological Change*, edited by Merritt Roe Smith, 39–86. Cambridge, MA: MIT Press.

Staub, Colin. 2017. "How Lightweighting Has Shaken up the Electronics Stream." *E-Scrap* (blog), May 4, 2017. https://resource-recycling.com/e-scrap/2017/05/04/lightweighting-shaken-electronics-stream/.

Staub, Colin. 2019. "Details on Total Reclaim Prison Sentences." April 25, 2019. https://resource-recycling.com/e-scrap/2019/04/25/details-on-total-reclaim-prison-sentences/.

Steinberger, Julia K., and J. Timmons Roberts. 2010. "From Constraint to Sufficiency: The Decoupling of Energy and Carbon from Human Needs, 1975–2005." *Ecological Economics*, Special Section: Ecological Distribution Conflicts 70, no. 2: 425–33. https://doi.org/10.1016/j.ecolecon.2010.09.014.

Strubell, Emma, Ananya Ganesh, and Andrew McCallum. 2019. "Energy and Policy Considerations for Deep Learning in NLP." In *Proceedings of the 57th Annual Meeting of the Association for Computational Linguistics*, edited by Anna Korhonen, David Traum, and Lluís Màrquez, 3645–50. Florence, Italy: Association for Computational Linguistics. https://doi.org/10.18653/v1/P19-1355.

TDI Sustainability. 2022. "Are You Ready to Shift from ESG to SDG?" August 2, 2022. https://tdi-sustainability.com/the-economist-thinks-esg-has-had-its-day-heres-why-i-agree/.

Thryft, Ann R. 2021. "3D Printing for More Circuits." Semiconductor Engineering, October 27, 2021. https://semiengineering.com/3d-printing-for-more-circuits/.

Torsello, Davide. 2012. "European Union." In *Encyclopedia of Consumption and Waste: The Social Science of Garbage*, edited by Carl Zimring and William Rathje, 242–44. Thousand Oaks, CA: SAGE Publications, Inc. https://doi.org/10.4135/9781452218526.

United Church of Christ Commission for Racial Justice. 1987. "Toxic Wastes and Race in the United States: A National Report on the Racial and Socio-economic Characteristics of Communities with Hazardous Waste Sites." Public Data Access.

United States Customs and Border Protection. 2013. "Drawback: A Refund for Certain Exports." https://www.cbp.gov/sites/default/files/assets/documents/2016-Dec/Drawback_refund_2%2812-16-2016%29_0.pdf.

United States Environmental Protection Agency. 1989. "Fairchild, Intel, and Raytheon Sites Middlefield/Ellis/Whisman (MEW) Study Area Mountain View, California: Record of Decision." https://www.nasa.gov/sites/default/files/atoms/files/record_of_decision_mew_study_area_june_1989.pdf.

United States Environmental Protection Agency. 2010. "Record of Decision Amendment for the Vapor Intrusion Pathway: Middlefield-Ellis-Whisman (MEW) Superfund Study Area Mountain View and Moffett Field, California." https://www.nasa.gov/sites/default/files/atoms/files/rod_amendment_mew_vi_august_2010.pdf.

United States Environmental Protection Agency. 2016. "Center for Corporate Climate Leadership Sector Spotlight: Electronics." Overviews and Factsheets. September 22, 2016. https://www.epa.gov/climateleadership/center-corporate-climate-leadership-sector-spotlight-electronics [Page no longer active] Archived page available at https://19january2017snapshot.epa.gov/climateleadership/center-corporate-climate-leadership-sector-spotlight-electronics_.html.

United States Environmental Protection Agency. 2023. "Metadata for Data Sources within PFAS Analytic Tools." https://echo.epa.gov/trends/pfas-tools.

United States Environmental Protection Agency. 2024. "Biden-Harris Administration Finalizes First-Ever National Drinking Water Standard to Protect 100M People from PFAS Pollution." News Release, April 9, 2024. https://www.epa.gov/newsreleases/biden-harris-administration-finalizes-first-ever-national-drinking-water-standard.

United States Environmental Protection Agency. n.d. "PFAS Analytic Tools | ECHO | US EPA." Accessed July 7, 2023. https://echo.epa.gov/trends/pfas-tools.

United States Geological Survey, and Thomas G. Goonan. 2005. "Flows of Selected Materials Associated with World Copper Smelting." Reston, VA: United States Geological Survey. https://pubs.usgs.gov/of/2004/1395/.

Valverde, Mariana. 2009. "Jurisdiction and Scale: Legal 'Technicalities' as Resources for Theory." *Social & Legal Studies* 18, no. 2: 139–57. https://doi.org/10.1177/0964663909103622.

Van der Velden, Maja. 2019. "Zombie Statistics and Poverty Porn." Africa Is a Country, March 28, 2019. https://africasacountry.com/2019/04/zombie-statistics-and-poverty-porn/.

Vogel, Jefim, and Jason Hickel. 2023. "Is Green Growth Happening? An Empirical Analysis of Achieved versus Paris-Compliant CO2–GDP Decoupling in High-Income Countries." *The Lancet Planetary Health* 7, no. 9: e759–69. https://doi.org/10.1016/S2542-5196(23)00174-2.

Woodruff, Jackson, David Schall, Michael F. P. O'Boyle, and Christopher Woodruff. 2023. "When Does Saving Power Save the Planet?" *HotCarbon '23: Proceedings of the 2nd Workshop on Sustainable Computer Systems.* New York: Association for Computing Machinery. https://doi.org/10.1145/3604930.3605719.

World Health Organization. 2021. "Children and Digital Dumpsites E-Waste Exposure and Child Health." https://apps.who.int/iris/rest/bitstreams/1350891/retrieve.

Wynne, Brian. 1987. *Risk Management and Hazardous Waste: Implementation and the Dialectics of Credibility.* London: Springer London, Limited.

Yang, Yongxiang, Rob Boom, Brijan Irion, Derk-Jan van Heerden, Pieter Kuiper, and Hans de Wit. 2012. "Recycling of Composite Materials." *Chemical Engineering and Processing: Process Intensification*, Delft Skyline Debate, 51 (January): 53–68. https://doi.org/10.1016/j.cep.2011.09.007.

3

Perspectives

This chapter presents five essays written by authors with direct professional and research experience on the topic of e-waste. The chapter opens with an essay by Dr. Ramzy Kahhat, a professor of engineering who has investigated a wide variety of topics associated with e-waste including international trade as well as informal repair and recycling economies in different national contexts. Dr. Kahhat situates e-waste from a perspective of the Global South. His essay provides an overview of the nuance and complexity with which end of life electronics need to be understood in context outside of richer nations of the Global North. Crucially, argues Dr. Kahhat, regulation of e-waste that ignores the diversity of Global South contexts risks imposing a policy environment that harms the very people and places it ostensibly is designed to protect from harm.

The next essay is by author and journalist Adam Minter who, like Dr. Kahhat, addresses misperceptions of e-waste as a global environmental problem. Minter's essay reflects on his experience in the early 2000s as a journalist reporting for trade magazines and business press covering the international scrap materials trade. In part, his essay focuses on how US- and UK-based NGOs and media turned a location in China called Guiyu into a symbol of environmental degradation caused by waste dumping. Minter's firsthand experience of the scrap trade and Guiyu led him to challenge the waste dumping narrative as overly simplistic. At the same time, he recounts how doing so also made it difficult to successfully pitch a more nuanced story about e-waste in China and elsewhere to media outlets beyond the more specialist scrap trade press.

Although reports from the media and other sources have often characterized the export of e-waste from wealthy nations to places like China and Ghana as "waste dumping," this narrative is overly simplistic and promulgates inaccurate stereotypes about the inhabitants of these countries. Economies benefiting from materials recovery and electronics repair connected to the import of post-consumer electronics tell a more nuanced story. (John New/Dreamstime.com)

104   *Electronic Waste*

The next two essays ask readers to reconsider the kinds of storylines that the authors, like Adam Minter, argue are overly simplistic representations of the people and places associated with end-of-life electronics. These two essays offer brief overviews of rich case studies of the creative work to recuperate and reuse discarded electronics by people in Accra, Ghana and Dar es Salam, Tanzania. The two sites are examples of places and the people who live and work there who, these authors argue, are all too often misrepresented in stereotypical tropes in popular media stories about e-waste as a global environmental problem. In the first of these two essays, Dr. Grace Akese describes the complex on the ground context of Agbogbloshie, a site in the capital city of Accra made notorious as an e-waste hotspot in a variety of international media stories about it. Dr. Akese's fieldwork at the site over the last decade details of vibrant materials recovery and electronics repair economy that is a key source of livelihood for people otherwise pushed to the margins. Her work also shows how international media representations of e-waste at Agbogbloshie have actually been appropriated by Ghanaian elites to usurp the land on which the site is situated for their own enrichment. In the second essay, Dr. Samwel Ntapanta recounts his fieldwork experience at *Mahakama ya friji* or the "Refrigerator Court" in Dar es Salam. Here, people engage in highly creative repair work that brings electronic devices, including refrigerators, into renewed and prolong useful lives for people in the city who would otherwise not be able to afford them. Both Akese and Ntapanta are clear eyed about the conditions under which people in these sites work. Those conditions, while exhibiting ingenious creativity, are also precarious and come with undeniable environmental and health risks. Like Kahhat and Minter, these authors argue that it is important to be mindful of unintended consequences that may arise from imposing remedies that do not meaningfully account for the conditions in these places and which may, unintentionally or otherwise, end up further marginalizing the people whose livelihoods depend on these activities.

The final essay of the chapter is by Dr. Melissa Gregg who has experience in both academia and industry. Dr. Gregg worked for more than a decade in user experience and sustainability research at Intel. Dr. Gregg's essay contrasts her involvement in researching computer users' "Out of the Box Experience" with a burgeoning industrial design area of practice on endings—that is, what happens when consumers are done with their electronic devices. Gregg describes existing and emerging approaches to take-back services offered by some major brands (e.g., Dell, HP, Lenovo, and Apple). Barriers to take back,

Perspectives 105

repair, and reuse are also covered. Gregg's essay concludes with forward-looking recommendations for addressing e-waste that pays better attention to upstream designs of digital devices that would enhance their longevity through repair and reuse.

The Light at the End-of-Life of Electronic Equipment: Narratives from the Global South

Ramzy Kahhat

The narrative of the end of life of electronics is often told under the context of the developed world or the cultural and socioeconomic context of the more affluent. However, the stories may differ from a Global South perspective or even within the less wealthy in the developed world. Thus, this essay intends to present the narrative of end-of-life electronics connected to the Global South.

Over the years, and especially in the Global North, electronic waste (e-waste) has been seen as a problem instead of an opportunity. The problem around the exportation of used electronics (sometimes referred to as e-waste, making invisible their capabilities for a second or even third life) to feed informal recycling practices in some parts of the world, such as China and India, as reported in the start of the twenty-first century, has been a standard in the narrative. While there are undeniable environmental and human health impacts related to the mishandling of e-waste in some parts of the world, such as open burning of copper cables, the use of methods such as acid baths to obtain precious metals from equipment, or other selective dismantling of e-waste and mismanagement of problematic parts or materials, there is also a different side that has received less attention in the media and from decision-makers. For example, some scholars have demonstrated that the trade of e-waste included equipment that pursued vibrant secondary markets in the Global South (e.g., used desktop and laptops; Kahhat and Williams 2009). In fact, this has been the case for many Global South countries, such as Peru, Mexico, and Nigeria (Estrada-Ayub and Kahhat 2014; Kahhat et al. 2008). Moreover, schools have also benefited from the trade of second-hand equipment or related initiatives, reducing the so-called digital divide, or technological gap between social groups, in many different areas of the globe (e.g., Brazil, Canada, and the United States).

In the Global South, classically, used products such as phones, laptops, and desktops tend to maintain their value, which is boosted due to the critical network of repair and maintenance (R&M) shops and informal technicians. R&M, reuse (including the reuse of parts), refurbishment, and repurposing are the most desirable paths at the end-of-use electronics. These are activities that have long been common for used electronics in the "developing" world, long before the more recent popularity of the idea of the so-called circular economy.

In a context where, traditionally, purchasing a new electronic device requires a savings plan or for low-income families is, commonly, unaffordable, it is not difficult to imagine the importance of secondary markets and the important work of maintenance, repair, and refurbishing that occur in those markets. Vibrant clusters of R&M businesses, often informal in character, are common in different Global South cities, such as downtown Lima, Peru and the cities of Mexicali and Mexico City, Mexico and elsewhere (Hieronymi, Kahhat, and Williams 2013; Kahhat et al. 2022; Lepawsky 2020). Indeed, during the last decade, R&M practices, including self-repair activities, have expanded in the Global North (Lepawsky 2020), changing consumer and social behavior in ways that were unimaginable 10–15 years ago to people outside of communities of highly technical repair hobbyists and technicians.

Aside from all the electronic life extension alternatives that prevail in the Global South, end-of-life of electronic devices also presents other economic opportunities and environmentally positive impacts if creative, flexible, and holistic thinking is used. For example, people who make a living collecting and processing the waste cast-off by others play a vital role in the Global South, including when it comes to the collection and dismantling of e-waste. This is of great importance in a context where formal waste management strategies are often insufficient, and formal recycling rates are low (Margallo et al. 2019). Also, it is important to note that the value of parts or materials found in electronics, established by the market, plays an essential role in the recovery activity. If there is a value for any particular part or material, the probability of it ending in a landfill or open dumpsite is heavily reduced by the labor of people making a livelihood in the informal sector. Creative solutions to provide value for these parts and materials are critical and it is often the informal sector that finds the most innovative solutions first. Clearly, it is important to recognize the role of the informal sector in waste management activities in the Global South and find ways to equitably support their activities while reducing the possibility of any environmental or human health risk related to their activities. A proper connection between their capabilities and those in the formal sector could enhance recovery of valuable parts and materials (Kahhat et al. 2022). This informal/formal linkage is a reality in many e-waste-related economies around the globe (Williams et al. 2013).

While there is a diversity of alternatives with regard to end-of-use of electronic devices, policies worldwide have typically ignored alternatives, briefly described in this essay, when dealing with their management. While it is easy to generalize and import successful strategies that ignore local contexts, proposals that ignore the unique characteristics of the system they intend to manage are prone to failure. Waste management proposals that ignore local contexts also risk negatively affecting the most vulnerable stakeholders of the system, such as people making a living as workers in informal economies. So, is there a light at the end-of-life of electronics? Yes, especially in those places where creative, flexible, and holistic strategies that consider the wide range of alternatives and opportunities at the end-of-use are carefully considered and put in place.

References

Estrada-Ayub, Jesús. A., and Ramzy Kahhat. 2014. "Decision Factors for E-Waste in Northern Mexico: To Waste or Trade." *Resources, Conservation and Recycling* 86 (May): 93–106. https://doi.org/10.1016/j.resconrec.2014.02.012.

Hieronymi, Klaus, Ramzy Kahhat, and Eric Williams, eds. 2013. *E-Waste Management: From Waste to Resource*. London: Routledge. https://doi.org/10.4324/9780203116456.

Kahhat, Ramzy, Junbeum Kim, Ming Xu, Braden Allenby, and Eric Williams. 2008. "Proposal for an E-Waste Management System for the United States." In *IEEE International Symposium on Electronics and the Environment*, 1–6. San Francisco. https://doi.org/10.1109/ISEE.2008.4562917.

Kahhat, Ramzy, and Eric Williams. 2009. "Product or Waste? Importation and End-of-life Processing of Computers in Peru." *Environmental Science and Technology* 43, no. 15: 6010–16.

Kahhat, Ramzy, T. Reed Miller, Sara Ojeda-Benitez, Samantha E. Cruz-Sotelo, Jorge Jauregui-Sesma, and Marco Gusukuma. 2022. "Proposal for Used Electronic Products Management in Mexicali." *Resources, Conservation & Recycling Advances* 13: 200065.

Lepawsky, Josh. 2020. "Planet of Fixers? Mapping the Middle Grounds of Independent and Do-It-Yourself Information and Communication Technology Maintenance and Repair." *Geo: Geography and Environment* 7, no. 1: e00086.

Margallo, María, Kurt Ziegler-Rodriguez, Ian Vázquez-Rowe, Rubén Aldaco, Ángel Irabien, and Ramzy Kahhat. 2019. "Enhancing Waste Management Strategies in Latin America under a Holistic Environmental Assessment Perspective: A Review for Policy Support." *Science of the Total Environment* 689 (November): 1255–75. https://doi.org/10.1016/j.scitotenv.2019.06.393.

Williams, Eric, Ramzy Kahhat, Magnus Bengtsson, Shiko Hayashi, Yasuhiko Hotta, and Yoshiaki Totoki. 2013. "Linking Informal and Formal Electronics Recycling via an Interface Organization." *Challenges* 4, no. 2: 136–53.

Ramzy Kahhat is Principal Professor at the Department of Engineering at Pontificia Universidad Católica del Perú. Kahhat obtained his Ph.D. and MSE in Civil and Environmental Engineering at Arizona State University. He is a broadly trained civil and environmental engineer applying concepts and methods from Sustainable Engineering, Industrial Ecology, and Earth Systems Engineering and Management. His expertise in these areas has been used in several research studies, such as waste and e-waste management, transboundary flows of e-waste, agro-industrial, urban and energy systems, urban stocks, and the characterization of debris generated by disasters.

Wasted Definitions: Negotiating Truth in Environmental Journalism

Adam Minter

It was 2002 when I first heard of Guiyu, China, the place that became, in name at least, the largest e-waste dumping site in the world. I was in my third month as a Shanghai-based correspondent for an American recycling trade journal. One afternoon a friend and I walked down to a local used consumer electronics market, one of tens of thousands that were sprinkled across China, offering used, imported US and European computers. I was in search of a laptop to replace one that had died. As I perused the vendors and their wares, I heard one tell another that he was awaiting a weekly shipment from "Guiyu." My friend stepped forward to ask if there would be desktops in the shipment, too. The vendor replied with a shake of the head. "No, only RAM and other chips."

That evening, in an odd coincidence, an American friend happened to email to me "The e-Wasteland," a story from *The Guardian* newspaper (Shabi 2002). It described Guiyu as a "grotesque, sci-fi fusion of technology and deprivation … where electrical waste from the west is routinely shipped for 'recycling.'" The reporting was primarily based on documentary evidence provided by an American activist group, which had created its own report, and which was also attached to the email. That document was filled with shocking photos of technology in flames, creeks and ponds clogged with plastic waste, young children posed amidst the detritus. In 2024, images like these have become so

Perspectives

109

commonplace that many news outlets and editors will turn down the opportunity to publish them out of boredom and ubiquity. That was not the case in 2002. Back then, they shocked.

As a trade journalist immersed in the Chinese scrap recycling trade, I'd long ago learned that the phrase "one man's trash is another man's treasure" isn't a mere aphorism; it's a business plan. To be sure, I'd seen a lot of trash, environmental degradation, and safety violations. But I'd also seen enough Chinese recycling, refurbishment, and reuse—including at my local electronics mall—to suspect that *The Guardian* report—and the activism that inspired it—were looking at waste and recycling in very different ways than the vendors at my local electronics mall. In China, and in the West, the definitions were continents apart. How to negotiate those definitions, it turned out, would become a major challenge of doing e-waste journalism.

In 2002, I accompanied a Chinese scrap metal dealer to Los Angeles scrap yards where he spent well over $1 million on a wild assortment of metal during a breakneck three-day business trip. Every day, I'd stand next to him as he looked into bins of metal. To my eyes, they contained a mix of wires, broken motors, and random bits and pieces of metals of various kinds. Who was going to separate this garbage out, much less recycle it profitably? To his eyes, they contained a rich potpourri of alloys that could be separated by hand in China and then sold at significant mark-up to manufacturers of new metal.

In other words: what I defined as trash, he defined as treasure. Over time, I adopted his definition, in part because it was the only way that I could explain to my readers why Chinese businesses imported millions of tons of recyclables every year. Put differently, nobody pays $1 million in three days for the privilege of shipping garbage to China. They do it for raw materials, and profits. That stuff isn't dumped; it's inspected, purchased, and imported.

But that's a tough idea to sell to jaded editors and readers. In my experience, there are two overarching and related reasons. First, over the last two decades, as anticolonial discourse has gone increasingly mainstream, so too has the ideological force of the "trash" definition in recycling discourse. Indeed, if you're only looking at the pictures, the global trade in waste looks like a kind of waste colonialism. "It's dumping, isn't it?" I was asked during a contentious email exchange with an editor at a prominent, national US magazine in 2005. "Why should we whitewash that with used computers?"

A second issue related to access. When I look at Western news coverage of Guiyu and other so-called e-waste zones, whether from 2002 or 2024, I'm struck by the fact that the reporting (video and photos, in particular) is collected almost exclusively outdoors. Think about it: how often have you seen a documentary about e-waste that includes shots from an indoor workshop? When have you seen an owner of an e-waste site submit to an interview, much less in her office? It might happen, but it's rare at best (Osseo-Asare and Abbas 2013). That's because the average business or environmental journalist assigned to do a story on e-waste doesn't have the time necessary to develop sources who'll take her inside the warehouse. She'll do her best, no doubt, but the best is—more often than not—a couple of weeks on the story before moving onto something else. Pressed for time, she'll visit Guiyu, take photos outside of the warehouses, and then write about what she saw outdoors.

That's a problem if the goal is to create an accurate picture of what's happening in an industry. You wouldn't, for example, write an article about the manufacture of automobiles by taking pictures of trash compactors in a factory parking lot or car shredding machine at an auto-scrapyard. You'd want to see the assembly line and talk to the people who designed it. It's same with e-waste. Precious metals, non-ferrous metals, and refurbished computer components— the marketable products of the e-waste industry—are high-value commodities that are extracted under controlled, indoor conditions, and stored in locked facilities. Nobody does the delicate work of extracting semiconductors from circuit boards under a blazing sun beside a riverbank. What makes sense on the riverbank, or in the adjoining fields, is the recycling and extraction of less valuable materials, like plastics, that don't require skill. It's dangerous, polluting work, and it's easy to photograph. But it's a revenue sideshow to what happens indoors.

It took me almost two years to cultivate a source who would take me inside Guiyu. This person, who asked to remain anonymous (and hasn't changed his mind in two decades) had gotten to know me over meals, and via my reporting on other aspects of the scrap trade. I believe he trusted me because he believed I had some understanding and sympathy for the complex supply chains required to recycle raw materials for Chinese business. In other words, he saw that I at least appreciated his definitions for waste and recycling; I wouldn't impose mine—at least, not on first glance.

So on a Friday morning we met at Hong Kong's airport, and drove the nearly six hours (it's much shorter these days) to Guiyu. I was prepared to be surprised, and I was. Guiyu wasn't a wasteland; it had a thriving downtown strip packed with restaurants and workshops with large signs advertising used semiconductors and memory chips. Inside the workshops I was introduced to

Perspectives

business owners who showed me the careful—and hazardous—process by which their employees lifted semiconductors off circuit boards so that they could be resold (the fumes would build up in those rooms, and become suffocating); they showed me rooms packed floor-to-ceiling with refurbished laptops, and safes packed with semiconductors extracted from boards that could be sold for hundreds of dollars for reuse. They also showed me their "waste" piles—circuit boards stripped of chips, hunks of plastic monitor shells—and drove me out of town to the places where it would be burnt and melted. The scenes were just as bad—and often worse—than what I witnessed in the documentaries and news coverage that had been reported in the area. Notably, one of the business owners revealed to me that most of the burning takes place after dark. Then he laughed and added: "We do it when the local government and the reporters have gone home."

Over the years, I found it difficult to place this reporting anywhere but trade magazines (and my books). Definitions, I found, were just too difficult to overcome. I'd seen the treasure; I'd also witnessed the trash. More important, I'd witnessed an environmental good—reuse and repair—amidst undeniable environmental harm. But nuance, whether applied to e-waste or any other field, doesn't generate many clicks. Equally important, it grinds against the consensus narrative about e-waste that took hold in the early 2000s. So long as news organizations are unwilling to devote organizational time and effort to develop a more complex understanding of the phenomenon, especially as it exists in emerging markets, that status quo will retain its power.

In fairness, there have been some improvements in the reporting environment since the Covid pandemic and a greater public interest in addressing the climate crisis. Recycling has fallen in status, and repair has risen. Major consumer brands are now embracing the idea and image (if not the actual reality) of repairable devices. As a result, reporters asked to cover the recycling of electronics (or any other consumer good) are less likely to assume that a broken computer is "trash" these days. That's an important step to a more accurate understanding of what "e-waste" means to the countries, businesses, and individuals that covet it.

References

Osseo-Asare, D. K., and Yasmine Abbas. 2013. "About Agbogbloshie Maker Space." *AMP* (blog), August 3, 2013. https://qamp.net/about/.

Shabi, Rachel. 2002. "The E-Waste Land." *The Guardian*, November 30, 2002, sec. Environment. https://www.theguardian.com/guardian/2002/nov/30/ weekendmagazine.pollution.

Adam Minter is an author and columnist who splits his time between Southeast Asia and the United States. He has written extensively on the global trade in recycling, including two books, Junkyard Planet: Travels in the Billion Dollar Trash Trade *and* Secondhand: Travels in the New Global Garage Sale. *In addition to writing about waste and recycling, he covers China, culture, and sports.*

A Closer Look at Agbogbloshie and the Electronic Waste Narrative

Grace Abena Akese

In this essay, I reflect on Agbogbloshie, a site in Accra, Ghana, notorious as a "hotspot" within the manifold spaces of global e-waste geographies. It is a site with which I have deeply engaged since beginning to do research there in 2010. In what contexts of global e-waste generation and movement is Agbogbloshie considered a hotspot? What significant developments have occurred in the almost two decades that the site has been known for its e-waste activities? Let's explore these questions.

If you happen to come across a news article about e-waste in the anglophone news media today, you will be told that Agbogbloshie is the "final resting place for e-waste from all over the world" (WIRED 2020). In stories like these, you, the reader—who is almost without exception presupposed to be a consumer in North America or Europe—will be asked to imagine that the desktop computer you donated to the recycling company in your city, if followed, will likely lead to Agbogbloshie, where workers, who are sometimes children, break devices apart with their bare hands under risky conditions. As the title of this article bluntly put it, "your old electronics are poisoning people at this toxic dump in Ghana" (WIRED 2020). This is the dominant narrative of the "e-waste problem" (see Lepawsky 2015). It is a tale that renders a world of e-waste producers in the Global North and victims in the Global South connected via a linear stream of e-waste flows.

Let's consider an alternative narrative, one that could not be ignored as I have been conducting fieldwork at Agbogbloshie and its region since 2010. Imagine

Perspectives 113

you are walking alongside me, tracing the journey of your discarded computer, but instead of ending up at Agbogbloshie, we take a different path. Where might we find your computer, and what could explain this unexpected detour? We might stumble upon your computer at one of the certified or regulated e-waste recyclers in your city. Surprising, isn't it? This recycler assures a secure chain of custody, carefully documenting the computer's whereabouts throughout the recycling process. They guarantee complete data erasure (you'll even receive a certificate of data destruction), break apart your computer, and separate it into different material streams like metals, plastics, and glass. I bet you'd be happy with this outcome despite the materials and energy lost in shredding your device (but let's bracket that discussion for now). You can pat yourself on the back for responsibly recycling your computer. You shouldn't feel targeted by the accusatory tone of that WIRED news article. But what this should tell you, and research has documented, is that the flow of discarded electronics isn't solely from the Global North to the Global South, as often portrayed in news media. The majority of e-waste flows are regional. This means your computer is more likely to circulate within your own country or even within regional blocks like NAFTA for North America or within Europe if you are part of the EU (see Lepawsky 2015).

If local or regional flows explain why your computer did not end up at Agbogbloshie, you may wonder why this place is so infamous for the e-waste crisis—as a problem of waste dumping. And where does the e-waste processed at Agbogbloshie come from? To understand this, let's take a quick look at the history of this site in the national and international context. Agbogbloshie began as a general scrapyard where broken-down vehicles used to transport food from Ghana's rural areas to the country's largest food market were repaired (Grant 2006). Over time, operations at the scrapyard expanded to include the sale of vehicle spare parts and the establishment of metal foundries that transformed scrap materials into various products. The site also began processing electronic waste, much of which was sourced locally from within Accra and elsewhere in Ghana. Ghanaians have been consuming electronics for decades, generating their own e-waste in addition to those arising from the import of second-hand electronics (see, e.g., Amoyaw-Osei et al. 2011). The country has one of the highest rates of mobile voice and data subscriptions, with an estimated 134 percent (meaning people have more than one mobile voice subscription) and 78 percent of the population, respectively (ITA 2023). Indeed, studies forecasted and are now showing rising domestic sources of e-waste in many developing countries, including Ghana (Lepawsky 2015; UNEP & Basel

Convention Secretariat 2012; Yu et al. 2010). The origin of e-waste processing at Agbogbloshie is therefore an example of the informal sector capitalizing on opportunities and building infrastructures "from the ground up" in the absence of state interventions (Silver 2014).

For years, the processing of e-waste at Agbogbloshie remained a low-key affair, receiving little attention both within Ghana and internationally. In Ghana, the scrapyard itself was seen as one of the many informal clusters that, through quiet encroachments, carve out spaces for themselves in a city that otherwise excludes the urban poor from access to land. Things took a turn in the late 2000s when the once obscured scrapyard suddenly generated widespread interest far outside Ghana and across the world. This increased attention was sparked by a pioneering toxicological study conducted by Greenpeace International (2008), an environmental non-governmental organization (ENGO). The study reported shockingly high levels of lead and cadmium pollution at the site, exceeding background levels by over 100 times. Greenpeace specifically attributed this alarming contamination to e-waste activities at the site yet did so without any substantive reference to the site's longer history as a scrapyard. Following the publication of Greenpeace's findings and other exposés about the site, Agbogbloshie quickly became a focal point for researchers, international journalists, development organizations, renowned photographers, and even "slum tourists" (tourists interested in marginalized or impoverished areas; for a more detailed history, see Akese 2019). As a result, Agbogbloshie underwent a rapid transformation, becoming an emblematic symbol representing the global e-waste crisis.

This spotlight on Agbogbloshie has had significant impacts, not only in terms of e-waste research and advocacy but also in driving a number of policies and interventions to address the risks associated with informal e-waste processing in Ghana. Prominent among them was the introduction of e-waste legislation in 2016, known as the Hazardous and Electronic Waste Control and Management Act 917. This legislation was accompanied by technical guidelines to provide more specific directions on sustainable e-waste management. Another intervention is a €25m funding partnership led by the German Development Organization (GIZ), which began in 2016 and will run until 2026 to promote "environmentally and socially responsible" management of e-waste in the country.

While these efforts led (or funded) by international development organizations strive to formalize the so-called informal e-waste sector that Agbogbloshie represents, broader urban transformation policies within Ghana exclude the sector and scrapyard workers from the urban space. Yet again, these broader

Perspectives   115

realities of social and economic exclusion of people pushed to the margins in Ghana are another tension relevant to a more nuanced and complex story about Agbogbloshie beyond the site's simplistic portrayal as the place where your computer goes to die. Agbogbloshie is part of a larger informal economy that faces violent evictions and community displacement in Accra. The site has been subjected to repeated demolitions and evictions, only to be rebuilt and occupied again (see Lepawsky and Akese 2015). In July 2021, as part of the "Let's Make Accra Work Again" urban transformation campaign led by municipal authorities in Accra, Agbogbloshie was razed down (Akese, Beisel, and Chasant 2022). Since then, scrapyard operations have relocated to the nearby old Fadama informal settlement, where most of the workers reside—a testament to the resilience and adaptability of informal scrapyard workers who continue to find ways to sustain their livelihoods amidst even dire hazardous working and living conditions.

Although Agbogbloshie has emerged as a symbolic hotspot within the wider landscape of e-waste geographies, the site is more than a place where your desktop computer goes to die. If you live in North America or Europe, your desktop is more likely to die in your regional "backyard" than make its way to Agbogbloshie. Like many informal clusters in rapidly urbanizing cities in Africa, places like Agbogbloshie rise to fill in gaps in infrastructural provision, although not without concerns for the health of people and environments. E-waste activities in Agbogbloshie are driven by a desire for economic opportunities, the absence of sufficient formal waste management systems and the growing generation of e-waste in a technology-dependent society in Ghana itself, as well as beyond it (see, e.g., Burrell 2012). I, therefore, urge you to rethink the Agbogbloshie you are often called on to imagine in mediagenic images and consider that, in its current predicament, this space highlights the complex and contentious nature of e-waste processing in a particular place and its intersection with broader urban development agendas within and beyond it.

References

Akese, Grace Abena. 2019. "Electronic Waste (e-waste) Science and Advocacy at Agbogbloshie: The Making and Effects of 'The World's Largest E-Waste Dump.'" Ph.D diss., Memorial University of Newfoundland. https://research.library.mun.ca/14273/.

Akese, Grace, Uli Beisel, and Muntaka Chasant. 2022. "Agbogbloshie: A Year after the Violent Demolition." *African Arguments* (blog), July 21, 2022. https://africanarguments.org/2022/07/agbogbloshie-a-year-after-the-violent-demolition/.

Amoyaw-Osei, Yaw, Obed Opoku Agyekum, John Pwamang, Esther Müller, Rapheal Fasko, and Mathias Schluep. 2011. "Ghana E-Waste Country Assessment: SBC E-Waste Africa Project." https://www.researchgate.net/publication/281381379_Ghana_e-waste_country_assessment.

Burrell, Jenna. 2012. *Invisible Users: Youth in the Internet Cafés of Urban Ghana.* Cambridge, MA: MIT Press.

Grant, Richard. 2006. "Out of Place? Global Citizens in Local Spaces: A Study of the Informal Settlements in the Korle Lagoon Environs in Accra, Ghana." *Urban Forum* 17: 1–24. http://www.springerlink.com/index/7J6YLGR3JY1TDK4X.pdf.

Greenpeace International. 2008. "Poisoning the Poor—Electronic Waste in Ghana." Greenpeace International. Accessed March 14, 2023. http://www.greenpeace.org/international/en/news/features/poisoning-the-poorelectroni/.

ITA, US Department of Commerce. 2023. "Ghana—Information and Communications Technology (ICT)." Accessed February 7, 2024. https://www.trade.gov/country-commercial-guides/ghana-information-and-communications-technology-ict.

Lepawsky, Josh. 2015. "The Changing Geography of Global Trade in Electronic Discards: Time to Rethink the E-Waste Problem." *The Geographical Journal* 181, no. 2: 147–59. https://doi.org/10.1111/geoj.12077.

Lepawsky, Josh, and Grace Akese. 2015. "Sweeping away Agbogbloshie. Again." *Discard Studies* (blog), June 23, 2015. https://discardstudies.com/2015/06/23/sweepingaway-agbogbloshie-again/.

Silver, Jonathan. 2014. "Incremental Infrastructures: Material Improvisation and Social Collaboration across Post-Colonial Accra." *Urban Geography* 35, no. 6: 788–804. https://doi.org/10.1080/02723638.2014.933605.

UNEP & Basel Convention Secretariat. 2012. "Where Are WEEE in Africa? Findings from the Basel Convention E-Waste Africa Programme." Accessed February 7, 2024. https://www.basel.int/Implementation/TechnicalAssistance/EWaste/EwasteAfricaProject/Publications/tabid/2553/Default.aspx#.

WIRED. 2020. "Your Old Electronics Are Poisoning People at This Toxic Dump in Ghana." *Wired.* Accessed June 7, 2024. https://www.wired.com/story/ghana-ewaste-dump-electronics/.

Yu, Jinglei, Eric Williams, Meiting Ju, and Yan Yang. 2010. "Forecasting Global Generation of Obsolete Personal Computers." *Environmental Science & Technology* 44, no. 9: 3232–37. https://doi.org/10.1021/es903350q

Grace Abena Akese is a geographer and discards studies scholar interested in waste geographies and political economies. She researches the geographies of electronic waste (e-waste) with a focus on Ghana.

Welcome to Solution? Repurposing of Electronic Waste in Dar es Salaam, Tanzania

Samwel Moses Ntapanta

Informal electronic waste recycling in the Global South, specifically sub-Saharan Africa, is considered an environmental and humanitarian crisis. This perspective is fueled on the one hand by media, policymakers, researchers, and international organizations like the European Union. On the other hand, e-waste is a new sphere for capital accumulation, for example when precious and semi-precious metals are recovered from circuit boards. Concerns about workers in the informal sector are often voiced with respect to the dangers of toxic compounds embedded in e-waste that informal workers are exposed to. Yet, exposure to e-waste toxic compounds also happens in the high-tech Global North recycling plants (Balasch et al. 2022; Julander et al. 2014). In such situations, people who work in the "informal economy" are often sidelined by people in more powerful positions. However, I see the work performed by people in the informal sector as offering solutions to the dangers of e-waste.

The Western media is filled with apocalyptic and dystopic images of e-waste recycling in Africa, especially those from the Agbogbloshie district in the West African city of Accra, Ghana. These kinds of stereotypes are well represented in the documentary "Welcome to Sodom" directed by Austrian filmmakers Florian Weigensamer and Christian Krönes, or *The Guardian* article "Life in Sodom and Gomorrah: The World's Largest Digital Dump" (Adjei 2014; Weigensamer and Krönes 2018). When I visited Dar es Salaam in Tanzania, one of the largest cities in East Africa in March 2018, I was also filled with such news and images of informal e-waste recycling. I searched for those places described in biblical terms of apocalypse, but they were nowhere to be found. I visited the *Pugu Kinyamwezi* landfill, the only landfill on the outskirts of Dar es Salaam. I hoped to see what I expected I would see. On the contrary, I was astonished by the small amount of e-waste that arrived there from the city of more than six million inhabitants. I asked myself, where was all the e-waste?

The Workshop

My quest to locate where e-waste ends up in Dar es Salaam led me to *Mahakama ya friji*. *Mahakama ya friji* (the refrigerator court) is an excellent case of the kind of ingenuity people have for making a living through the creative reuse of equipment others may have cast-away. The court is located in the middle of *Kinondoni*, a bustling low-income residential area known as "Uswahilini," in the middle of Dar es Salaam. A few buildings hide *Uswahilini* from passers-by on the busy Kawawa road 300 meters away. The location is strategically chosen to avoid being plainly visible to government authorities but near to sources of e-scrap and customers of the products. Among the micro-enterprises here, there are around twenty-five to thirty crafters who make charcoal stoves out scrap metal. There are also around twenty scrap collectors, most in their late teens or early 20s, and three scrap dealers, the actors in between the collectors, manufacturing companies, and exporters who send reclaimed materials and components to destinations beyond Tanzania.

The name *Mahakama ya friji* reflects the activities at this place. Defunct materials collected by scrap collectors around the city are brought here. Upon arriving at the workshop, they are valorized and salvaged. Repairers buy parts that can still be used to fix other broken but still usable devices. Crafters purchase aluminum boards to produce charcoal stoves, kitchenware, and roofing materials. The rest of the materials, including palladium, copper, iron, and cast are snapped up by scrap dealers who send them to export markets outside of Tanzania.

The Refrigerator

Rama, one of the collectors with whom I went along on many collecting trips, is a master in valuing defunct devices. He told me once, "If you do not know how to valuate, do not become a collector." On two of our trips, I observed his skills to know almost the exact worthiness of a device. On both trips, a refrigerator was presented to us by the owner, who wanted to discard it. One was an old Bosch, and the second, though defunct, looked like a recent LG design. Rama negotiated and paid a low price for the LG and a higher for the Bosch. A few months later, I asked him why he paid low for the newer LG? He told me that the older Bosch, especially the compressor, is preferred by repairers who replace it with some of the new types, mainly from China. And even if the compressor is beyond repair, he told me Bosch is made for the European market. That means the amount of copper in the compressor weighs more than in the new LG. More copper means higher scrap metal value.

Once returned to the workshop, the compressor from Bosch was bought right away by a repairer. The aluminum boards were gone to the stove crafters, and

Perspectives 119

Rama made more money than buyers would pay for the newer LG model. The LG compressor was dismantled to mine the copper because repairers didn't see the use of it. Collecting e-waste is a complex endeavor that requires mastery of skills and knowledge to identify, negotiate over, and price goods. When a device is presented to a collector, they have to know very quickly what kind and weight of metals are in that device, otherwise a potential seller may pass on their offer. They have to negotiate a buying price with the seller while at the same time keeping in mind what price they will get from repairers, crafters, and scrap dealers. All of these meticulous calculations happen in a matter of a few minutes. These knowledge and skills do not nearly resemble the simplistic biblical images of hell and damnation used by the Western media to characterize places like *Mahakama ya friji*.

The Stove

"Hey Sam, let us check for aluminum sheets to make your coffee friend a nice stove," George invited me after I had shown eagerness to make a cookstove. I had befriended a coffee brewer who has a coffee shop near my residence in Dar es Salaam. After observing him struggling to make coffee and ginger tea on a small charcoal stove, I promised to make a double charcoal stove for his business: one side for coffee and the other for ginger tea.

George and I walk among the pushcarts, looking for aluminum boards from refrigerators for my coffee stove. After carefully inspecting all refrigerator and air conditioners, he returns to one refrigerator before saying, "This one is just perfect." Then, George negotiated the price with the collector. They settled at 20000 TSH (about US$8) without the compressor and the wiring inside. Calvin (an apprentice with George) and I dragged the refrigerator to the dismantling space. We started by hammering the door out, then the back and the sides, to get the copper wires out. After about thirty minutes, we had the aluminum dismantled from the insulators.

As we arrived at the workshop the following day, George was already there waiting for us. He instructed Calvin and me to smooth the aluminum sheet while he was finishing the drawing for the stove. We pulled our rail hammers under a tree and hammered the aluminum bends. George came with the drawing as soon as we finished smoothing. He instructed us to cut the different parts of the stoves following the measurements written on the drawing while he bent iron rods for the stove's coils. I took the easy job of producing the legs of the stoves. This is because I was a new apprentice without fully developed skills. Calvin produced the rest of the stove. George kept checking if we were following the instructions and being precise with our measurements.

When our parts were ready, George started stitching and welding them together. After two hours, the stove was ready. While we were painting the stove, which is usually done with black paint, George went inside the workshop to pick two clay pots to fit on the two holes for the charcoal. The final stage was to install the coils, and the stove was ready. I delivered it to my friend, the coffee brewer, that evening.

What my experiences doing field research in Dar es Salaam taught me is to view with skepticism claims that electronics reuse, repair, and recycling in countries like Tanzania are inherently "primitive" and of lesser value than those in the United States or elsewhere. Simplistic narratives that usually bypass practices I have described incite politics over the danger or toxicity of e-waste in developing countries without also critically examining recycling in developed countries that also expose workers to toxic chemicals (Ntapanta, forthcoming).

Conclusion

In the current technological epoch, we believe that only advanced technology will solve our problems. Meanwhile, micro-scale, ingenious activities like those at *Mahakama ya friji* should be noticed, instead of ignored or marginalized. These unrecognized activities that spontaneously arise out of these unregulated patches create platforms for a potential sustainable, if more "informal", urbanization. These "micro activities" assemblages are valuable in building the city and the world. They do not merely compensate for missing technology and infrastructure (Simone 2004). They are practices that envisage a future through ingenuity and creativity that gradually create grounds for the solutions people are looking for. As researchers with expectations founded on what we retrospectively understand to be our own taken-for-granted assumptions, the thing to do is to stretch our gaze to see this world in our backyard that gathers immense potential for the prosperity of our planet.

References

Adjei, Asare. 2014. "Life in Sodom and Gomorrah: The World's Largest Digital Dump." *The Guardian*, April 29, 2014, sec. Working in Development. https://www.theguardian.com/global-development-professionals-network/2014/apr/29/agbogbloshie-accra-ghana-largest-ewaste-dump.

Balasch, A., M. López, C. Reche, M. Viana, T. Moreno, and E. Eljarrat. 2022. "Exposure of E-Waste Dismantlers from a Formal Recycling Facility in Spain to Inhalable Organophosphate and Halogenated Flame Retardants." *Chemosphere* 294: 133775. doi: 10.1016/j.chemosphere.2022.133775.

Julander, Anneli, Lennart Lundgren, Lizbet Skare, Margaretha Grandér, Brita Palm, Marie Vahter, and Carola Lidén. 2014. "Formal Recycling of E-Waste Leads to Increased Exposure to Toxic Metals: An Occupational Exposure Study from Sweden." *Environment International* 73 (December): 243–51. doi: 10.1016/j.envint.2014.07.006.

Ntapanta, Samwel M. (forthcoming).*Gathering the African Technosphere: Informal E-Waste Recycling in Tanzania*. Lanham, MD: Lexington Books.

Simone, Abdou Maliqalim. 2004. "People as Infrastructure: Intersecting Fragments in Johannesburg." *Public Culture* 16, no. 3: 407–29.

Weigensamer, Florian, and Christian Krönes, dirs. 2018. *Welcome to Sodom*. Documentary. Blackbox Film & Medienproduktion GmbH. https://www.welcome-to-sodom.com/.

Samwel Moses Ntapanta is an ethnographer of contemporary urbanism along the western Indian Ocean coast (East Africa), particularly on coloniality, consumption and discarding, debris of late capitalism, repairing and recycling economies.

Designing for Reuse

Melissa Gregg

The Reuser Experience

When I first joined the user experience research team at a large technology company, one of the methods used to study consumers' attitudes to computers was the "Out of Box Experience"—the "OOB." This technique involves following what happens from the moment the box is opened and the product inside is successfully set up for first use. The researcher's task is to note any difficulties encountered in the process to feed back to design and engineering for improvement.

The OOB method pre-dates popular social media genres now dedicated to "unboxing." Both practices reflect the excitement and hope that attends the beginning of a product relationship. As sustainable design advocate Joe Macleod (2017, 171) argues, by optimizing the seamless commencement of these relationships with technologies, "we have almost completely overlooked how to stop them. The digital industry is in denial about endings." This essay shares examples of designers focused on this problem of endings.

When a product reaches "end of life," as the business world calls it, reuse is the most environmentally welcome outcome to reduce the environmental impact of its production. In the North American context, among other advanced economies, Original Equipment Manufacturers (OEMs) such as Dell, HP, Lenovo, and Apple provide take-back services for products that are no longer wanted, whether through mail-back costs, drop-off destinations, even doorstep pickups in certain neighborhoods. These "reverse logistics" solutions allow products to be resold, either through digital platforms or wholesalers that operate domestically and internationally. Retailers are also investing in this "re-commerce" trade, which became popular when fresh hardware stocks depleted during the Covid pandemic.

The process of disassembling and refurbishing electronics at scale in commercial transactions is known as "IT asset disposition." When a leasing term expires for a company's computer fleet, for example, the ITAD's in-house experts audit the quality and value of the aging devices. Valuable parts are harvested to sell or shred for reuse in other applications. Each machine's stored data are cleared using strict standards enforced by accrediting agencies. A system comprised of reassembled parts can provide critical online access to those in need, such as the populations served by nonprofit organizations addressing the digital divide.

In other settings, device repair and reuse is more distributed and entrepreneurial. Local cultures of repair and resale offer work-arounds to official channels, especially when hardware brands engage in restrictive design and distribution (Hernández-Tapia 2023). In both formal and informal markets, the labor cost of repair can make this work hard to justify. But this is only because the environmental costs of mining, manufacturing, assembling, and shipping new products have not been factored into pricing historically (Grossman 2010).

Resisting Hibernation

One of the biggest barriers to reuse is the volume of idle electronics in homes and workplaces. The Circular Electronics Partnership (cep2030.org) states that "80% of e-waste is not collected for recycling with 76% not documented." Researchers call this prolonged state of device inactivity "electronics hibernation" (Snyder and Rooks 2021). It presents an issue because the longer such devices are out of circulation, the shorter the time window for them to be given to another owner.

Interviews reveal that people hold on to their devices for a range of reasons. They don't know how or where to dispose electronics locally, for instance, or they want a backup in case the new device fails. Often, it's a problem of time. Important files and photos may need sorting into more permanent storage locations. The transfer costs of moving data around can seem overwhelming and just boring. For those who can already afford to upgrade, there isn't much financial incentive to profit from reuse.

The fact remains that when digital objects are left idle or in storage, reuse becomes harder. The software needed to keep hardware running requires constant maintenance and may fall out of date after too long. In our research at Intel, even when people did try to log in to an old device, there were often unforeseen problems. A crucial connecting cable was missing, or a forgotten password rendered the system useless.

The software required to facilitate hardware reuse is a growing concern. With large companies like Microsoft, Samsung, and Apple deciding the terms on which repair technicians can operate, it is becoming harder to run open-source alternatives to cloud-dependent computing services. Software and hardware increasingly work together in the same company's interests, as was the case when Google acquired a startup to monetize the reuse of Windows PCs and Chromebooks (Wheatley 2020).

Designing Products for Reuse

Product stewardship, circular design, and cradle-to-cradle thinking are some of the common design approaches tackling reuse (see, e.g., McDonough and Braungart 2002). Emphasis is placed on making best use of the resources embodied in a product, through repurposing, gifting, recycling, and more. The cradle-to-cradle concept is a tweak on industry terminology supporting "cradle to grave" product development. The implication is that companies should take responsibility for the product experience through each "life stage," anthropomorphizing the object. But it is increasingly apparent that the waste problem exists because there has not been enough emphasis on product endings. No rituals have developed to ensure technology users find closure, to gracefully finish the device relationship and "consciously uncouple."

It used to be much easier to fix electronics and use them for longer. Early laptop designs made it simple to swap an old battery for a new one, for example, or upgrade storage, which is still the case with desktop computers today. But as more people wanted to carry a laptop from place to place, or use mobile phones, engineers needed to build differently. Smaller devices used smaller screws, while soldering and adhesives helped them stay secure and protected from damage. As devices became lighter, the trade-off was that parts became harder to upgrade. The only way to access new features was to replace the whole system. This history is a popular way that OEMs explain their resistance to changing reference designs, the "recipes" that make certain ways of building the industry standard (Gregg 2023). New entrants to the hardware ecosystem in recent years have tested whether these claims hold weight.

In at least two prominent examples, modular design is making a comeback. In Europe, Fairphone was the first modular smart phone designed with hardware longevity in mind. The Framework laptop, from the San Francisco-based company of the same name, takes this premise further: the system's modular design means the components can be separated and selectively upgraded depending on user preference. Framework calls it "the last computer you'll ever buy" because the system is intended to last. Users become builders, putting together precisely the computer they want. They also become sellers by trading components through the online Framework Marketplace. This approach to technology reuse may be intimidating for those who already find technology confusing, or lack the confidence to buy and sell goods online. In the context of the broader industry, however, it is a rare development to suggest that consumers buy less.

As the name implies, the "personal computer" was built for the benefit of the individual while minimizing the social and environmental impacts of its production (Maxwell and Miller 2012). A combination of activism, regulation, and business innovation is consolidating to produce a better future for design informed by the collective costs of consumption. Ultimately, to address e-waste will require deeper reflection on the experience with which this essay began: to ask why so many now experience intimacy with new technology as a mark of accomplishment, as entertainment, and a source of sublime fascination.

References

Circular Electronics Partnership. https://cep2030.org/.
Gregg, Melissa. 2023. "Talking in Circles about E-Waste." June 2023. https://melgregg.com/2023/06/14/talking-in-circles-about-e-waste/.

Perspectives

Grossman, Elizabeth. 2010. *Tackling High-Tech Trash: The E-Waste Explosion & What We Can Do about It*. New York: Demos. https://www.demos.org/press-release/new-report-examines-high-tech-trash-e-waste-ahead-black-friday.

Hernández-Tapia, Lidia. 2023. "The Genius Bar in a Country Where iPhones Can't Legally Be Sold." Rest of World, October 2023. https://restofworld.org/author/lidia-hernandez-tapia/.

Macleod, Joe. 2017. *Ends: Why We Overlook Endings for Humans, Products, Services and Digital. And Why We Shouldn't*. UK: Joe Macleod.

Maxwell, Richard, and Toby Miller. 2012. *Greening the Media*. New York: Oxford University Press.

McDonough, William, and Michael Braungart. 2002. *Cradle to Cradle: Remaking the Way We Make Things*. New York: North Point Press.

Snyder, Carrie, and John Rooks. 2021. "Breaking up with Our Products: Encouraging Electronics Product Dis-attachment to Enable Secondary Uses." Product Lifetimes and the Environment (PLATE) Conference, May 26–28, 2021. https://hdl.handle.net/10344/10195.

Wheatley, Mike. 2020. "Google Buys Neverware, Whose Software Turns Old PCs into Chromebooks." Silicon Angle, December 16, 2020. https://siliconangle.com/2020/12/16/google-buys-neverware-whose-software-turns-old-pcs-chromebooks.

Melissa Gregg is the author of over sixty books and articles on the future of work, technology, and everyday life, including Work's Intimacy *(2011) and* Counterproductive *(2018). For ten years, she worked as Senior Principal Engineer in user experience research and sustainability at Intel.*

4

Profiles

There are hundreds of governmental and NGOs around the world focused on various aspects of pollution and waste arising from electronics. Many, though not all, direct their activities to e-waste as an end-of-pipe problem, that is, downstream as waste that arises when consumers get rid of devices they no longer want. A smaller number of these organizations direct their activity upstream toward the mining for, and manufacturing of, electronics which represent the stages of the lifecycle of electronics where most of the pollution and waste arise. Although this chapter is not exhaustive, it attempts to provide an overview of some of the key actors that have and continue to play dominant roles in shaping approaches to managing e-waste internationally. In that respect, there is a predominance of European and North American-based groups. This is partly because organizations in these regions were among the first to profile e-waste as a special category of environmental concern. Profiled here are ENGOs, trade associations, national government agencies and intergovernmental organizations, certification bodies, and a variety of nonprofit groups.

Agbogbloshie Maker Space

The Agbogbloshie Maker Space (AMS) is an ongoing project organized by architects and designers based in Ghana, France, and the United States. Originally, the project sought to build semi-permanent and mobile workshop

---

Chatham House, home to the Royal Institute of International Affairs, in London, UK. The British think-tank was founded in 1920 to research matters of international concern. It supports circulareconomy.earth, an online portal of publicly available data about the international trade of primary and secondary materials, including statistics about transboundary flows of post-consumer e-waste. (Chris Dorney/Dreamstime.com)

spaces for people who live and work in what was the Agbogbloshie scrap yard in Accra, Ghana. The Agbogbloshie scrapyard was a site with a long history of negative representation of it by a variety of local and extra-local interests. In the mid-2000s it became infamous in European and North American-based media as a site of e-waste dumping. The AMS project approached the site and the people who live in work there from a different angle. The project sought to create workspaces with relevant occupational health and safety protections and locally appropriate equipment for creative work using materials derived from the scrapyard, including post-consumer e-waste.

Over the years, Agbogbloshie had been a site of repeated so-called slum clearance by Ghanaian authorities, but in July 2021 the site was definitively cleared to make way for real estate development. AMP has since reoriented its activities to design modular and mobile workspaces using lighter materials, such as locally available bamboo. These structures can be built more quickly in place especially in the dispersed sites of electronic scrap refurbishment, repair, reuse, and recycling that have since emerged after the demolition of Agbogbloshie. The AMS website is a useful resource for e-waste researchers looking for ways to think creatively about how post-consumer discarded electronics can be creatively reappropriated and worked with in safe(r) conditions.

American Chamber of Commerce in Europe

The American Chamber of Commerce in Europe (AMCHAM EU) advocates for American companies that do business in the EU. Its priorities focus on advocating for trade, investment, and business competitiveness within the framework of a growth-based economy between the European Union and the United States. AMCHAM EU lobbies European parliamentary members to shape the agenda of the European Council in favor of the business interests AMCHAM EU represents. The organization's membership includes many companies directly related to electronics manufacturing and retailing such as Amazon, Apple, Broadcom, Google, HP, IBM, Intel, Meta, Microsoft, Oracle, and Qualcom, among others.

AMCHAM EU's policy priorities are wide ranging. They cover agriculture and food, competition policy, consumer affairs, customs and trade facilitation, the digital economy, environment, financial services, healthcare, intellectual property, security, defense and space, taxation, trade and external affairs, and transport, energy, and climate. Electronics are an important part of all of these economic sectors.

The organization's remit is broad, but it has played specific roles in shaping how the Basel Convention deals with e-waste. AMCHAM EU played a direct role in the years-long negotiations over technical guidelines intended to distinguish between waste and non-waste electronics under the Basel Convention. During these negotiations the AMCHAM EU has advocated for exemptions for specific device categories from their designation as waste electronics under the guidelines. In particular, AMCHAM EU made the case for exemptions for medical devices and industrial electronic as distinct from consumer electronics. AMCHAM EU argues that discarded medical and industrial electronics should be exempt from restrictions on transboundary movements since they often require specialized analysis to determine reasons for failure and repair services that are not always available in all countries where such devices are in use.

Bamako Convention

The Bamako Convention is an international treaty amongst twenty-five African countries that prohibits the import and transboundary movement of hazardous waste, including e-waste, to or between parties of the convention. Bamako and the Basel Convention are directly related to one another. Both treaties entered into force in the 1990s in response to notorious incidences of waste dumping in African countries from wealthy nations of the "Global North." They also largely mirror one another in their definitions of hazardous waste. Yet, the Bamako Convention exhibits some notable differences from the Basel Convention. For example, the Bamako Convention contains regulatory language that specifically invokes clean production and that what counts as such shall explicitly exclude end-of-pipe pollution controls. In other words, Bamako seeks to mitigate and eliminate pollution and waste upstream in production before products are made in the first place. In this respect, the Bamako Convention offers a more impactful regulatory model than other forms of regulation that treat e-waste as an end-of-pipe problem that arises after devices have already been manufactured.

Basel Action Network

The Basel Action Network (BAN) is a nonprofit ENGO founded in 1997 and based in Seattle, Washington. It operates internationally. The organization advocates in support of the United Nations' Convention on the Transboundary Movements of Hazardous Wastes and Their Disposal (i.e., The Basel Convention). BAN's advocacy is focused on four campaigns: advocacy around the Basel Convention, e-waste stewardship, ship recycling, and the international plastic waste trade. In 2009, the group created a third-party e-waste recycling certification standard called e-Stewards to distinguish electronics recycling businesses that handle such recycling in a manner deemed safe under e-Stewards criteria.

BAN published an important report called *Exporting Harm: The High-Tech Trashing of Asia* in 2002 that profiled the then emerging issue of exports of discarded electronics from the United States to China, India, and Pakistan. That report touched off much of the academic, journalistic, and broader public interest in the trade and traffic of post-consumer e-waste. The findings of this report and other work by BAN have played a direct role in justifying bills introduced in the US Congress dealing with exports of electronic waste from the country. BAN's research has been explicitly referenced in some of the bills brought to the floor of Congress although none of these bills have yet passed into law.

The organization has historically played, and continues to play, a key role in negotiating amendments to the Basel Convention. Since 1999, the organization has played recurring roles in contributing to various technical working groups. This work covers a wide range of relevant issues under the Basel Convention. Early advocacy by BAN made recommendations for improving medical waste management under the Convention as well as improving the handling of ship scrapping, how legal liability is handled, linking the Basel Convention two other major conventions under the United Nations environment program—the Rotterdam Convention (covering hazardous chemicals and pesticides) and the Stockholm Convention (covering persistent organic pollutants). Beginning in 2010 BAN began playing a central role in shaping the technical guidelines on electronic waste under the Basel Convention. Those technical guidelines remain in interim use in part because BAN and other actors who are part of the negotiations remain concerned that provisions for repair of discarded electronics offer a loophole in the guidelines that would allow bad actors to export e-waste by claiming the export is for repair.

Basel Convention

The Basel Convention on the Control of Transboundary Movement of Hazardous Wastes and Their Disposal (hereafter, the Basel Convention) was originally adopted in 1989. It came into force on May 5, 1992. There are 191 parties to the treaty and 53 full signatories as of 2024, a notable exception being the United States which has signed, but not ratified the Convention.

The Basel Convention arose out of growing international concerns about the consequences of the tightening of hazardous waste legislation in many industrialized countries beginning in the 1980s. A key concern was that those tightened regulations would lead to a growth in international trade in hazardous waste for disposal in poorer countries. Infamous cases of such toxic trading did indeed occur in the early to late 1980s. The Basel Convention aims to regulate such international trade such that when it occurs it involves prior notification of intention to export and prior consent to receive an import by parties who agreed to trade. Furthermore, the convention requires that transboundary shipments of hazardous waste result in the treatment of that waste through environmentally sound processing.

The wording, meaning, and interpretation of the Basel Convention have been a source of disagreement between different coalitions of actors since its adoption in 1989. Coalitions of different states, trade associations, and NGOs have battled over language and definitions of what will count as hazardous waste as well as how those definitions may or may not conflict with different national-level regulations, and business interests involved in the international trade of scrap materials for recycling. One acute area of disagreement was over a particular annex of the convention known as Annex VII. An Initiative dubbed the "Basel Ban" was championed in 1994 by members of what would become the Basel Action Network (BAN). BAN and coalition partners, particularly African states, advocated for a complete ban of hazardous waste exports from countries defined under Annex VII (member states of the Organization for Economic Cooperation and Development [OECD], the European Community [EC], and Liechtenstein) to all non-Annex VII countries. Annex VII and the Basel Ban divided up the parties of the Basel Convention into so-called developed and developing nations and sought to prohibit the trade of hazardous waste from the former to the latter. Advocacy around the Basel Ban continued for years. Some nations, including some "developing" countries and certain business interests resisted the adoption of the ban because they thought it would deprive them of a source of revenue derived from managing, processing, and reselling recycled materials back to global commodity markets. Those interests resistant to the adoption of Annex VII managed to stave off its inclusion in the Basel Convention until the late 2010s. Eventually, at a general meeting of parties to the convention the Basel Ban was adopted and came into force in December 2019.

Other important points of ongoing debate and negotiation remain in the current version of the Basel Convention in general regarding hazardous waste and e-waste specifically. For example, Annex IX covers categories of wastes that will be excluded from the meaning of hazardous waste in other parts of the convention. One of the categories in this annex is electrical and electronic assemblies intended for direct reuse, but not for recycling or final disposal. The Convention then goes on to describe reuse as inclusive of repair, refurbishment, and upgrading, but not what it refers to as major reassembly. Unfortunately, the Convention remains silent on what counts as "major reassembly," and therefore creates loopholes that bad actors could take advantage of, at least potentially. These kinds of ambiguities or loopholes have been a major impetus for attempts to better define what will count as waste versus non-waste electronics under the Convention.

Negotiations over the crafting of such technical guidelines began in earnest in 2010. As in negotiations over other parts of the convention, the crafting of these guidelines has gone on for years. By 2015, a draft set of guidelines had emerged but were only adopted on a provisional and interim basis. As such, while the guidelines provided some guidance on what would count as waste versus non-waste electronics, there was no legally enforceable provision in the Convention that made that distinction. The most recent round of negotiations on these technical guidelines was published in March of 2023. This draft of the guidelines remains adopted on an interim basis. There remain many points of contention within the guidelines in terms of how they attempt to distinguish waste from non-waste electronics. Providing exemptions for export for failure analysis and repair remains a particularly contentious issue. Those favoring such an exemption continue to point out that advanced failure analysis and repair infrastructure do not exist in every country where device brands are sold and used. Those opposing such exemptions continue to see them as a dangerous loophole that bad actors could use to export waste electronics for activities otherwise prohibited by the Convention such as recycling and disposal.

Profiles 133

Bureau of International Recycling

The Bureau of International Recycling (BIR) is a nonprofit trade association representing the interests of the international recycling industry. BIR was founded in 1948 and is headquartered in Brussels, Belgium. The organization currently represents over 900 individual companies in seventy countries. BIR's work covers eight commodity categories: ferrous metals, nonferrous metals, paper, textiles, steel and special alloys, plastics, tires and rubber, and electronic scrap.

BIR cultivates an extensive network of contacts with international bodies to influence the global legislative landscape regulating the international recycling trade. It represents its members' interests in national legislative bodies as well as a variety of international bodies such as the UN, OECD, and the EU. This advocacy includes lobbying policymakers on legislation that regulates the recycling industry. The advocacy on the part of BIR takes the form of in person communications with policymakers, the publication of market data and research, and position papers on a variety of topics relevant to the recycling industry and the commodity categories it represents. An important part of BIR's advocacy around electronics recycling has been to rework the idea of e-waste away from the negative connotations of scrap and waste and toward a more positive understanding of this category as a resource material.

Center for Public Environmental Oversight

The Center for Public Environmental Oversight (CPEO) is an ENGO based in Mountain View, California. It originally formed in 1992 in response to a growing number of closures of military bases in the region that left behind contaminated sites. However, the organization's roots also trace to another nonprofit public interest research group called the Pacific Studies Center (PSC) formed in the 1970s. PSC was an occasional meeting place for organizers who would go on to form the Santa Clara Center for Occupational Safety and Health (SCCOSH; see separate profile).

The work of PSC and CPEO is focused primarily on the Silicon Valley region and the impacts of toxicants on people and places in the region arising from the electronics sector and its connections to military contractors and bases in

134 *Electronic Waste*

the region. In 1977, the PSC published one of the earliest critical examinations of the high-technology industry in Silicon Valley (called *Silicon Valley: Paradise or Paradox?*). This highly prescient report details many problems associated with the electronics industry in the region that have only grown more acute over time and remain relevant today. For example, the report documents the deep connection between foundational companies such as Fairchild Semiconductors and the broader military-industrial complex in Silicon Valley. It also covers early negative consequences on housing affordability and commute times. The report provides details of negative labor conditions and occupational health and safety conditions for workers in the sector differentiated by gender and race. It also points to early signals of environmental contamination in the region by the electronics sector.

The work begun at PSC has been carried on at CPEO. The organization continues to produce important research, commentary, and legal briefs regarding labor conditions and toxicants in the electronics sector within and beyond the Silicon Valley region.

With the passage of the US CHIPS legislation, which aims to strengthen US manufacturing, supply chains, and national security by investing in research and development and science and technology, CPEO has more recently turned its attention to the labor conditions, occupational health and safety issues, and environmental consequences of toxicants in use by, and emitted from, the semiconductor industry. Some of the organization's most recent reports provide information on semiconductor manufacturing and greenhouse gas production, land-use change, and specific concerns around "forever chemicals" (e.g., PFAS). Research and reports going back decades are free to download from CPEO's website. The work of CPEO is critical to broadening the understanding of pollution and waste associated with electronics beyond what consumers discard.

Circular Electronics Partnership

The Circular Electronics Partnership (CEP) is an industry advocacy organization dedicated to enacting and electronics sector premised on circular economy principles. The organization is comprised of founding partners, a Secretariat, and regular members. CEP's founding members include the Global Electronics Council (GEC; see separate profile), the Global Enabling Sustainability Initiative (GeSI; see separate profile), the International Telecommunication

Union (ITU; see separate profile), the Responsible Business Alliance (RBA), the World Business Council for Sustainable Development (WBCSD), and the World Economic Forum (WEF). The secretariat is comprised of mining and recycling companies, many well-known brand electronics manufacturers, retailers, and software enterprises.

CEP coordinates the range of projects geared toward implementing a circular electronics industry envisioned by its members. The organization has produced what it calls the Circular Electronics Roadmap and the Circular Electronics System Map, both of which are publicly available on the organization's website. Each document presents high-level overviews of barriers to and possibilities for implementing business models premised on circularity in the electronics sector. As such, the roadmap and the system map pay attention to issues such as acquiring resources, design, manufacturing, collection of post-consumer discards, and the reincorporation of materials back into manufacturing supply chains. In addition to these high-level documents, CEP also links to several high-level reports on issues associated with a circular electronics system. From an e-waste researcher perspective, CEP offers insight into how powerful players in the mining for, and manufacturing of, electronics understand pollution and waste from the industry beyond the narrow focus of post-consumer discards.

Commission for Environmental Cooperation

The Commission for Environmental Cooperation (CEC) is an inter-governmental body created between Canada, Mexico, and the United States under the original North American Free Trade Agreement (NAFTA) in 1994. Its secretariat is based in Montreal, Quebec. The CEC is intended to facilitate inter-governmental cooperation on issues of environmental conservation and protection. It is comprised of a governing Council, a Secretariat, and a Joint Public Advisory Committee (JPAC), the latter being composed of fifteen regular citizens of five each from Canada, Mexico, and the United States.

The CEC is an important source of data on pollution and waste arising from the electronics industry. Companies operating within the NAFTA region are required to report to the CEC releases of any of 659 substances identified as hazardous. These reports are compiled by the CEC into a pollution release and transfer registry (PRTR) that is made publicly available online and contains data on pollution releases from approximately 40,000 individual industrial facilities across the NAFTA region.

Initially, reporting to the CEC was a voluntary measure, but since 2004 reporting has become mandatory in all NAFTA countries (Canada, Mexico, and the United States). CEC makes its PRTR data publicly available through its "Taking Stock" online portal and annual report. "Taking Stock" is a user-friendly database freely available to the public that reports the handling of hazardous substances at the individual facility level across industrial sectors of the NAFTA economy, including the manufacturing of electronics. The database allows users to separate out facilities by economic sector (e.g., electronics manufacturing) and find data on hazardous pollution arising at individual facilities within a given sector. Additionally, the database offers the ability to map the location of individual facilities across a given economic sector giving users the ability to create maps of industry sector specific pollution across the NAFTA region.

The data available through the CEC come with important caveats. First, they are incomplete in that the data cover only a selective list of substances deemed hazardous rather than all hazardous substances released by industry. The data also exclude all non-hazardous waste. Consequently, the CEC data do not cover total waste arising. Other important caveats include that each NAFTA country has historically had different regulatory thresholds for reporting pollutant releases from industry. To some extent these differences are harmonized under NAFTA, but their differences mean that analysis of trends over time can be difficult. However, while certainly not perfect, the CEC data offer an important window on pollution and waste arising upstream in the industrial phase of electronics production.

Chatham House

Chatham House is the informal name given to the Royal Institute of International Affairs, a British think tank originally established in 1920 and located in London, England. Chatham House was founded to research matters of international concern and the organization has become one of the most important independent think tanks devoted to international affairs. Indeed, its flagship journal *International Affairs* is among the top peer-reviewed journals in the field of international relations.

Today the organization operates as a not-for-profit entity. It is supported by both public and private sources of funds. Major government supporters include the Foreign, Commonwealth and Development Office of the UK and Global Affairs Canada. Various private corporations also donate to Chatham House, including electronics firms such as Apple and Meta.

Chatham House established *circulareconomy.earth*, an online portal of publicly available data about the international trade of primary and secondary materials. Secondary materials are those derived from recycling, and these Chatham House data include statistics about transboundary flows of post-consumer e-waste. Users of the database can map international flows of e-waste overtime as well as download data for their own analytical purposes.

The Chatham House data are derived from International Merchandise Trade Statistics (IMTS) that are collected under the auspice is of national customs authorities. These data are fed into a United Nations trade database called the Commodity Trade Statistics Database (UN Comtrade). IMTS and UN Comtrade are not particularly user-friendly and part of the impetus for Chatham House to create its *circulareconomy.earth* web-portal was to offer a more broadly accessible way to access and analyze these data. Chatham House researchers have developed methods for overcoming or mitigating data gaps and errors arising from the original data sources (i.e., IMTS and UN Comtrade). The result is a publicly available, user-friendly web-based interface for accessing, mapping, and downloading data for a variety of purposes, including investigations of the international trade of post-consumer waste over time.

DigitalEurope

DigitalEurope is a trade association founded in 1999 and headquartered in Brussels, Belgium. The organization is the outgrowth of an amalgamation of several industry associations associated with information and communications technology, telecommunications, and the consumer electronics manufacturing and retailing industries. Like other lobbying organizations, DigitalEurope advocates to policymakers on behalf of the industries the organization represents. Principally, this means advocating for legislation and regulation that will enhance (or at least not harm) the operating environment for the organization's membership.

DigitalEurope has played an important role in negotiations over the technical guidelines for distinguishing between waste and non-waste electronics under the Basel Convention. In particular, the organization argued that it would not be economical to have original equipment manufacturing repair available in every country where different digital devices are sold and used. Consequently, DigitalEurope advocated for exemptions under the guidelines that would allow for malfunctioning equipment to be exported for failure analysis and repair. Other actors in these negotiations consider such exemptions for repair to be a loophole that can be exploited too easily by bad actors to export post-consumer waste for disposal.

Electronic Products Recycling Association

The Electronic Products Recycling Association (EPRA) is a not-for-profit producer responsibility organization (PRO) that organizes end-of-life electronics collection and recycling in Canada's provinces and territories. PROs have become the preferred organizing mechanism for many jurisdictions' extended producer responsibility (EPR) systems. PROs typically function by collecting and distributing the environmental handling fees charged to consumers who purchase new electronics. These fees are used to fund electronics recycling systems.

EPRA originally formed in 2003 when it was called Electronics Product Stewardship Canada (EPSC). Its goal then was to develop a national electronics recycling program in the country. Major brands and retailers formed the original membership of EPRA. They were concerned about being faced with a patchwork of electronics recycling legislation that might differ province by province across Canada. The organization formed and wrote model legislation that has shaped all but one of the provincial and territorial electronics take back programs that now exist (the exception is Alberta's).

For e-waste researchers EPRA's website provides a potentially useful source of information and data from each province's and territory's electronics recycling programs. Researchers can find detailed annual reports from each jurisdiction as well as a variety of education and marketing material.

Electronics Watch

Electronics Watch is an innovative civil society organization based in Amsterdam. The organization is focused on protecting and enhancing the rights of workers in global electronics manufacturing supply chains. The organization advocates for public procurement policies in support of workers' rights including occupational health and safety on the factory floor.

Governments represent one of the single largest purchasers of electronic devices. Consequently, public procurement policy is a lever for enhancing workers' rights in the electronics sector including around occupational health and safety (OHS). One key area of OHS is limiting or minimizing exposure to toxic chemicals in the workplace. In this sense, Electronics Watch plays an important role in reducing or eliminating upstream releases of pollution and waste in the electronics sector.

Electronics Watch focuses its work on three main areas: capacity building for public authorities interested in using their purchasing power to enhance worker rights in the electronics sector, worker-driven monitoring, and engagement with industry. Electronics Watch supports capacity building of public authorities by raising awareness about labor conditions in global electronics supply chains. The organization also provides support for supply chain mapping and assists in finding ways to incorporate human rights due diligence directly into public procurement planning and contracts. The organization facilitates worker-driven monitoring by partnering with other civil society organizations in locations hosting electronics production facilities as well as mining regions. Electronics Watch helps coordinate and train workers to engage in monitoring and auditing strategies. Evidence gathered from the organizations monitoring activities is used to inform multi-stakeholder dialogues between public procurement officials, industry representatives, and workers with the goal of enhancing workers' rights, including occupational health and safety, throughout the electronics supply chain from manufacturing back to mining.

Electronics Watch produces publicly available guides and studies as well as monitoring reports and policy briefs all focused on working conditions in the mining for, and manufacturing of, electronics. This makes the organization a key source of information on those phases in the lifecycle of electronics where most of the pollution and waste arising from them occur.

E-Scrap News

E-Scrap News is an essential specialized source for industry journalism focused on the e-waste recycling industry in North America. It is part of a suite of industry journals published by Resource Recycling Inc that cover recycling news

140 *Electronic Waste*

and policy. E-Scrap News offers a wide range of useful information and data for researchers interested in e-waste. Running topics covered by the site include keeping track of changing certification standards, collating and analyzing lawsuits, the shifting discussions about e-waste exports, how markets for recyclable materials derived from electronic scrap are changing, and summaries of new research relevant to the industry. Researchers will find links to primary documents, including court cases and data sources. Although E-Scrap News focuses on the North American industry, it also provides good coverage of developments in Asia and Europe. Readers have the option to subscribe to a print version of E-Scrap News, but the online edition is free to access and past editions of the print magazine are archived digitally and free to read online. There is also a free weekly email digest available for readers.

E-Stewards

E-Stewards is a certification system for electronics recycling originally developed by the Basel Action Network (BAN). Electronics recyclers who wish to be certified under the e-Stewards program are subject to an initial auditing process to gain certification and ongoing auditing requirements to maintain their certification. BAN developed the e-Stewards program partly in response to what the organization saw as shortcomings of a years-long, multi-stakeholder process originally facilitated by the US EPA to develop an independent certified electronics recycling system for US-based processors.

E-Stewards is designed to adhere to regulations promulgated under the Basel Convention. The program relies on a system of interconnected not-for-profit accreditation organizations that authorize private for-profit auditing firms that, once qualified, can provide auditing services for e-Stewards and other third-party electronics recycling certification systems. Like other certification systems discussed in this chapter, e-Stewards represents a form of private, market-based regulation of the international electronics recycling trade.

Eurostat

Eurostat is the statistical authority of the European Union. Its mandate is to produce high-quality statistics on a wide range of topics relevant to all EU countries. The offices of Eurostat are physically located in Luxembourg City, Luxembourg, and the data collected by the organization is available online. First established in 1953, Eurostat has evolved over the last seven decades into a comprehensive statistical agency with multiple roles in the EU. These roles include data coordination, communication and dissemination, and data quality assurance.

Eurostat is a key source of data for e-waste researchers. For example, researchers can access data on waste electrical and electronic equipment (WEEE) for the EU as a whole and for any of its individual member states. Eurostat defines WEEE via EU Directive 19/2012 on waste electrical and electronic equipment promulgated in July 2012. Under these regulations, statistics on WEEE refer to any substance or object which a holder disposes of or is required to dispose of under national law and which is dependent on electricity below a define threshold to function. A helpful feature of Eurostat data is that statistics on a given topic are accompanied by short, plain language articles that provide further context to the data as well as links to relevant legislation and related concepts and data about those topics. These features are clearly organized and permit learning and analysis accessible to both novice and advanced statistical researchers. The Eurostat data on WEEE links to associated concepts such as recycling and waste recovery as well as relevant legislation governing WEEE in the EU.

Other data sets available from Eurostat that are useful for e-waste researchers include transboundary shipments of waste by sending and receiving countries. These data permit the tracking of where shipments across borders within and beyond the EU. These data categorize each shipment by a hazardous characteristics code, the amount shipped stated in metric tons, the country or countries of transit through which waste shipments move, the country of origin and destination, a categorization of the specific disposal operation, and a description of the final recovery operation associated with each waste shipment.

Another useful data set covers businesses in the computer and personal household goods repair sector. Eurostat is one of the few international statistical agencies that provide data specific to consumer electronics repair services. As such it provides a quantitative lens on electronics repair an important waste diversion strategy. These data also provide important insights into the employment structure and business size of the sector and how they differ across EU regions.

Researchers with advanced experience and research needs can access additional documentation covering highly detailed technical analyses of the

142   *Electronic Waste*

methods used by Eurostat to collect and publish data on particular topics. Eurostat offers a trustworthy source of information for researchers by providing access to clear basic and advanced documentation about the data it provides.

FreeGeek

FreeGeek is a US-based not-for-profit organization founded on Earth Day in the year 2000 in Portland, Oregon. Initially the organization focused on recycling and reuse of unwanted electronics for the purpose of waste diversion from landfills. The organization accepts donations of used equipment from businesses, organizations, and individuals. It uses these donations for its repair, refurbishment, and recycling activities.

Since 2000, FreeGeek has enlarged its activities to cover digital inclusion through the provision of free education related to digital technology, access to computing resources, and digital literacy. Today the organization is primarily focused on narrowing digital divides within the communities in which it operates. Recovery, refurbishment, repair, and recycling remain core activities and provide the hardware with which FreeGeek offers its community services related to digital education and literacy.

The organization offers several innovative services. One is "Plug into Portland," a program to promote access to digital technology for students in Oregon's K-12 school system. The organization notes that as many of 75,000 students in Oregon do not have a personal computer to complete their online assignments. FreeGeek helps overcome this gap by providing schools and students with hardware derived from the repair and refurbishment activity done by the organization. Another service offered by FreeGeek is a hardware grant program. Organizations of various sizes can apply to receive free refurbished digital technology for use in qualifying nonprofit and community organization activities. A third service offered by the organization is the sale of used electronic devices repaired and refurbished by FreeGeek volunteers. The organization sells these items online and point out that as many as 27 percent of Americans do not own a computer and that one in ten families do not have access to the internet at home. As such, this repair and refurbishment for resale activity undertaken by FreeGeek is an important way of overcoming various digital divides. It also offers people who may not be able to afford the cost of brand-new electronics an

Profiles 143

affordable way to access digital hardware. FreeGeek's approach also offers a way for consumers looking to reduce their personal environmental impacts related to electronics a way to purchase devices that helps conserve the materials and energy embodied in those devices and reduce the pollution and waste associated with their production.

Global Electronics Council

The Global Electronics Council (GEC) is a not-for-profit trade association based in Portland, Oregon. It originally formed as the Green Electronics Council in 2005. The organization's primary goal is to use the power of public procurement (i.e., government purchasing) to positively impact the design and manufacturing of electronics. Public procurement can be a powerful market influencer because of the magnitude of its purchasing power.

GEC members worked to create the Electronic Product Environmental Assessment tool (EPEAT). EPEAT evaluates electronics devices on a wide range of environmental criteria, partly derived from the Institute of Electrical and Electronics Engineers (IEEE) own standards for sustainability (i.e., IEEE Standard 1680). After the GEC's initial work the EPEAT standard was released in 2006. Since then, EPEAT has been officially incorporated into public procurement requirements of the US federal government. Under successive executive orders signed first by George W. Bush and then Barack Obama, US federal agencies are required to purchase electronics that meet EPEAT standards. The presence of such a significant market participant means that electronics brands that wish to sell to the US government are given a significant incentive to design and manufacture their devices so that they meet EPEAT standards. The EPEAT standard is an important intermediary into the pollution and waste impacts of the electronics manufacturing industry. In this respect, it is an important intervention into the broader meaning of e-waste beyond a narrow focus on post-consumer discards.

The GEC website offers e-waste researchers a wide range of useful information and data. For example, the organization keeps up-to-date review of the state of sustainability research in a variety of electronic device categories. There are also purchasing guides available for public procurement professionals as well as case studies, white papers, and related industry news.

Global Enabling Sustainability Initiative

The Global Enabling Sustainability Initiative (GESI) is an industry advocacy organization dedicated to applying digital technology to sustainability issues as well as to shifting industry practices toward sustainability within the global electronics supply chain. Its membership is comprised of well-known electronics manufacturers, telecommunications companies, international accounting and auditing firms, internet platforms, and associated business interests. The work of the organization is mostly focused on the application of digital technology to sustainability issues. However, it is also a member of other industry advocacy organizations such as the Circular Electronics Partnership (CEP) which focus on questions of environmental impact associated with the mining for, and manufacturing of electronics. GESI does pay some attention to pollution and waste arising from electronics manufacturing, mostly in the form of emissions from industry with a particular focus on the sector's implications for the climate emergency.

Global E-waste Statistics Partnership

The Global E-waste Statistics Partnership (GESP) is a nongovernmental policy actor organized by the International Telecommunications Union (ITU), the United Nations University (UNU), the United Nations Institute for Training and Research (UNITAR), and the International Solid Waste Association (ISWA). The GESP is focused on developing standardized statistical methods for improving and collecting e-waste statistics from countries around the world. The partnership publishes continually updated data on e-waste arising via periodic reports covering the global picture as well as regional snapshots. It also produces individual country reports and an online map that provide data on several facets of e-waste generation. These include total annual new electrical and electronic equipment sold in the national market, total mass e-waste arising, e-waste arising per capita, data on the formal collection of e-waste as well as measures of e-waste imports and exports.

GESP argues that standardized methods for e-waste data collection and reporting offer a crucial policy tool. For example, they enable cross-country comparisons that can identify countries doing more or less well to properly manage e-waste. GESP data enable the tracking of trends over time and the setting and assessment of targets for e-waste minimization and elimination. These data can also be used by countries looking to assess their contribution to international projects such as the United Nation's sustainable development goals. GESP also provides capacity-building workshops for national statistical agencies around the world. These workshops are directed to professional statisticians and help with understanding the principles of developing national-level e-waste statistics within each nation's own statistical agency that are comparable to others around the world.

The GESP website is a crucial resource for e-waste researchers. Beyond the organization's own data and map, researchers can access dozens of reports and other publications from both GESP and associated organizations. Since GESP began its work in 2017 it has become a key actor in the global e-waste management discussion. The organization's Global E-waste Monitor Report, published every two years, offers a gold standard of e-waste data.

Good Electronics Network

The Good Electronics Network is a nonprofit civil society organization based in the Netherlands and focused on human rights and environmental sustainability issues in global electronics supply chains. It is composed of a coalition of trade unions, other grassroots organizations, research organizations, academics, and activists. Good Electronics bases its work on the UN's Guiding Principles on Business and Human Rights. The organization argues that businesses have a duty to protect against human rights abuses and environmental harm in their supply chains. The organization favors legally binding government regulation on workers' rights, corporate accountability, and environmental protection.

Good Electronics focuses its activity on several topics that address pollution and waste from electronics beyond the narrow focus on post-consumer e-waste. One of these topics is chemical exposure. The organization notes that global electronics supply chains rely on large volumes of hazardous chemicals, especially in the manufacturing process. Good Electronics advocates for workers' right to know about the toxic hazards they may be exposed to in the workplace. Many of the hazardous chemicals used in the manufacturing process are not incorporated into the final manufactured device. Instead, they are used in intermediate processes by which those devices are created and can make their way into workers' bodies and the broader environment through various exposure and emissions pathways. Good Electronics and its coalition partners provide research and monitoring reports that point to key gaps in how companies handle hazardous chemicals and worker exposures to them.

Researchers can freely access over 150 reports on Good Electronics' website covering the full range of topics addressed by the organization. Researchers can also find well-produced videos on labor conditions in electronics supply chains including both the mining for, and manufacturing of, these devices.

Greenpeace

Greenpeace is an international ENGO formed in 1969 and currently headquartered in Amsterdam. It originated in Vancouver, British Columbia, Canada, with the Don't Make a Wave Committee, a group of people advocating against nuclear weapons testing in the Aleutian Islands of what is today Alaska. Since its founding Greenpeace has branched into activities broadly concerned with environmental issues and peace activities. It is among the most important and influential international ENGOs in the world and attracts praise and criticism concomitant with that status. Greenpeace's advocacy tactics are the subject of substantial analysis, support, and criticism. The group is noted especially for its media-savvy actions used to garner and focus broad public attention onto matters of concern to the organization.

Among its key priorities is advocating against toxic waste emissions and pollution by industry. It has focused some of its advocacy about this issue on the global fashion and textile industry as well as the global electronics industry. Greenpeace has produced several reports and advocacy materials about post-consumer e-waste, toxicants in electronic devices, and releases of toxicants during their manufacture. For example, the organization produced a *Guide to Greener Electronics* for several years. These reports ranked electronics brand manufacturers devices using a variety of criteria to provide consumers with a list of more and less environmentally friendly electronics. In this respect, the campaign was premised on the idea of enhancing consumer awareness to change their shopping behavior toward more environmentally friendly electronic devices. A contemporaneous program called *Clicking Clean* ranked internet companies by the degree to which they used renewable electricity to run their data centers. Both the *Guide to Greener Electronics* and the *Clicking Clean* report programs came to an end in 2017. Greenpeace has also produced several influential reports on post-consumer e-waste sites in Africa and Asia.

Profiles 147

iFixit

iFixit is a privately held company headquartered in San Luis Obispo, California. The company sells tools and parts for repairing consumer electronics and other devices. It also provides free, step-by-step repair guides for a wide range of products. The guides are primarily focused on electronics but also include manuals for repair of apparel, vehicles, medical devices, appliances, and households. At the time of writing there are over 106,000 free manuals covering over 56,000 devices available on the iFixit website. The plethora of manuals is created by iFixit staff, large numbers of volunteers, and by educational institutions that partner with the business. The creation of repair guides is reviewed to ensure they meet iFixit's standards for quality and completeness. The review system is partly managed via reputation scores earned by contributors when they comment on or edit existing guides in ways that the user community finds useful and rewards with reputational points.

iFixit supports educational resources and advocates for right to repair legislation both in its home country of the United States and around the world. People wishing to learn skills related to technical writing can follow a free, step-by-step online course by signing up with a school email address and becoming an iFixit a community member. iFixit is a co-founder of repair.org, which lobbies for right-to-repair legislation in the United States. An important impetus behind the company's right-to-repair advocacy is the concern about the volume of post-consumer e-waste arising in the United States and around the world. One of iFixit's key concerns is changing laws that prohibit independent and do-it-yourself (DIY) repair activities. The company argues that restrictions on repair violate other existing rights related to ownership as well as contributing to the growing volume of post-consumer e-waste. Their argument is that right to repair enhances peoples' abilities to conserve the materials and energy embodied in their devices and, thus, reduce pollution and waste arising from them.

Information Technology Industry Council

Information Technology Industry Council (ITI) is a trade association with offices in Washington, DC, and Brussels, Belgium. ITI's membership includes many well-known brands of electronics and software. The organization advocates for the interests of its membership across a wide range of fields relevant to information and communication technologies. What is today ITI

148 *Electronic Waste*

originally formed in 1916 in Chicago, Illinois, as the National Association of Office Appliance Manufacturers. Since then, the organization has gone through periodic change as new information technology categories and applications became relevant. ITI's overarching goal in all its areas of advocacy is to create a public policy and regulatory environment that is amenable to the business goals of its members.

Among its multiple areas of policy focus is environment and sustainability. In this regard ITI contributes to policy discussions related to post-consumer e-waste and the Basel Convention, including directly participating in negotiations over the technical guidelines distinguishing waste and non-waste electronics. In those negotiations ITI is among the actors advocating for exemptions for certain categories of equipment that would allow for export for repair and reuse. The organization's website offers e-waste researchers access to several ITI policy documents. These documents layout in detail ITI's position on policy and regulation in various jurisdictions around the world regarding environment and sustainability issues, including e-waste, as well as the organization's other priorities for lobbying.

International Campaign for Responsible Technology

The International Campaign for Responsible Technology (ICRT) is a network of organizations advocating for social, environmental, and economic justice in the global electronics industry. ICRT includes community groups, workers organizations, and consumer groups in approximately fifteen countries relevant to the electronics sector. ICRT activities cover the full lifecycle of electronics from mining, to manufacturing, to use, to discarding. The organization supports campaigns against specific high-profile electronics companies such as Apple and Samsung. It does so in part because of the overall market power of these brands. Campaigns that lead to shifts in practices that align with ICRT's goals have a high probability of becoming adopted more broadly across the sector due to the market position of these high-profile brands.

ICRT has history spanning four decades. Its origins go back to 1978 and the work of the Santa Clara Center for Occupational Safety and Health (SCCOSH; see separate profile). In this respect, ICRT is an important organization for its focus on pollution and waste arising in manufacturing of electronics, among its other priorities. It also distinguishes the organization from other ENGOs concerned with e-waste but with a focus mostly on post-consumer discards. SCCOSH activists and allies would go on to form other important activist groups such as the Silicon Valley Toxics Coalition (SVTC; see separate profile).

An early campaign of the ICRT centered on a cancer cluster associated with a National Semiconductor manufacturing site in Scotland. The organization also worked in coalition with NGOs based in Europe who were advocating for new legislation covering waste electrical and electronic equipment (WEEE) in the region. This work represented an internationalization of the original Campaign for Responsible Technology (CRT) that formed in 1990 and the group revised its name to reflect this.

In the early 2000s, ICRT's advocacy Network expanded to places in Asia, such as Taiwan, as the electronics sector itself increasingly globalized. In 2006, the organization supported the publication of an edited book called *Challenging the Chip*. The book contains research and essays from researchers and activists associated with the work of ICRT in multiple countries. The book covers topics related to labor conditions, pollution, and waste throughout the electronics lifecycle. It remains a key reference work for research into pollution and waste arising from the electronics sector.

More recently ICRT has expanded its advocacy in relation to the United States CHIPS Act. This act represents a monumental shift in industrial policy for the United States as an attempt to reshore key segments of electronics manufacturing supply chains, especially semiconductors, after decades of offshoring of that activity, especially to Asia. Many of the concerns that originally sparked ICRT's advocacy remain relevant to this newly shifting policy landscape, including occupational health and safety for workers as well as environmental justice for people and communities that host existing and new facilities impacted by CHIPS funding.

International Solid Waste Association

The International Solid Waste Association (ISWA) is a nonprofit trade association based in Rotterdam, the Netherlands. ISWA advocates for policy and legislation that will benefit the business operations of its membership. The organization co-sponsors a variety of initiatives related to making publicly available e-waste awareness campaigns, reports, and statistics. For example, ISWA is a co-sponsor

150   *Electronic Waste*

with International Telecommunications Union (ITU) of the Global E-waste Statistics Partnership (GESP) (see separate profiles).

ISWA supports a peer-reviewed journal called *Waste Management and Research* aimed at waste management industry professionals, scientists, engineers, and government managers. The organization also supports a magazine called *Waste Management World* which offers news and opinion about the international waste management and recycling industries. Both publications contain many articles relating to e-waste and its management.

International Telecommunications Union

The International Telecommunications Union (ITU) is an international organization formed in 1865 and based in Geneva, Switzerland. In the late 1940s, it folded into the United Nations system. The work of ITU is dedicated to developing technical standards that enable telecommunication systems of different jurisdictions to interoperate with one another. Since the days of the telegraph, the remit of the organization has expanded to include standardizing digital network communications protocols as well as coordinating the allocation of global radio spectrum and satellite orbits.

ITU has several key areas of action that include access to information and communication technology (ICT), artificial intelligence, broadband, cyber security, and the environment and climate change among other areas. Post-consumer e-waste is one of ITU's key concerns under its environment and climate change umbrella. The organization is a sponsor of the *Global E-waste Monitor*. It also collates research on e-waste and periodically publishes its own data on the subject. In addition, ITU offers several free learning resources about e-waste accessible online.

Interpol

The International Criminal Police Organization (Interpol) is an international body that facilitates the cooperation of police forces around the world. The organization's roots are found in a 1923 conference held in Vienna, Austria. Today, Interpol's ambit covers all major forms of crime, including what the organization calls environmental crime.

Interpol has been involved in multiple investigations of alleged e-waste trafficking over the years. It has also initiated its own programs and investigations into the issue. For example, in the mid-2010s Interpol supported "Project Eden" and the Countering WEEE Illegal Trade (CWIT) project. Project Eden arose largely out of concerns about e-waste pollution in west Africa, principally Ghana and Agbogbloshie. The initiative sought to raise awareness, establish what Interpol calls National Environmental Security Task Forces (NESTs), enhance the capacities of enforcement agencies, and suppress criminal activity. The Interpol-sponsored CWIT project resulted in an important report that provides quantitative evidence against the most common understandings of e-waste dumping. While that study did find illegal exports of WEEE from European countries to African destinations, it also found that ten times more e-waste was being illegally traded within the European region itself.

Interpol's activities around e-waste were most prominent between approximately 2009 and 2015. The organization's website suggests that its most recent activities on the issue were in 2017. Some reports, images, and other potential sources of data on the illegal trading traffic of e-waste remain publicly available on Interpol's website.

Oeko-Institut

The Oeko-Institut is a nonprofit research association based in Germany and founded in 1977. Its original focus was research in support of sustainable development at the local, national, and international levels. Today, the institute employees almost 400 staff and researchers who collaborate on research projects involving natural scientists, economists, social scientists, engineers, and lawyers. Since its founding, its areas of expertise have grown to include topics such as climate policy, transition pathways for energy, employment, mobility, and environmental law. The institute also continues to focus on issues of digitization including research on aspects of post-consumer discarded electronics.

Oeko-Institut researchers worked in conjunction with the Basel Convention and the United Nations Environmental Program, among other groups, to produce some of the early research on e-waste exports to Africa and e-waste arising domestically on the continent. In 2014, the institute produced an

152   *Electronic Waste*

important report called *Global Circular Economy of Strategic Metals—the Best-of-Two-Worlds Approach.* The latter report became the basis of policy discussions about how e-waste recycling systems based in richer countries could work in collaboration with those in poorer countries to achieve outcomes beneficial to both regions without compromising the health and safety of workers or the environment. The Oeko-Institut continues to produce research and reports relevant to e-waste issues. These are freely available from the Institute's website.

Open Repair Alliance

The Open Repair Alliance is a network of organizations and volunteers advocating for more durable and more repairable electrical and electronic devices. The Alliance was founded by five organizations dedicated to repair: the Anstiftung Foundation based in Germany, Fixit Clinic and iFixit (see separate profile) both based in the United States, the Repair Café Foundation (see separate profile) based in the Netherlands, and the Restart Project (see separate profile) based in the United Kingdom.

The Open Repair Alliance helps coordinate a global "Repair Day" held annually in October. Repair Day events occur around the world and a variety of groups circulate their repair activities through social media and in person community events. The Open Repair Alliance distinguishes itself from other volunteer repair advocacy groups through a focus on data standardization, data collection, and storytelling.

Creating an open standard for repair data is an important and challenging project. Consider earlier discussions in this book about how different jurisdictions categorize electronics to be covered under e-waste legislation, for example. Repair data faces some of the same challenges. The Open Repair Alliance's work on data standardization means that a growing database of internationally comparable information on electronics repair is publicly available. The data are available under a Creative Commons license and can be downloaded for free from the Open Repair Alliance website. At the time of writing, the data set contains over 133,000 records of repair. These data contain information about the brand, model, and year of manufacture of a device, the attempted repair and its outcome, where the repair took place, and what community group was involved.

Organisation for Economic Co-operation and Development

The Organisation for Economic Co-operation and Development (OECD) is an intergovernmental organization established in the aftermath of the Second World War. As the Organisation for European Economic Co-operation (OEEC), it was originally comprised of the United States and western European nations solely for coordinating postwar reconstruction under the Marshall Plan. After years of negotiations, the OEEC was transformed into the OECD in 1961. The OECD's main goal is facilitating economic growth and financial stability through the expansion of world trade.

The OECD's relevance to e-waste comes in multiple forms. The important one is the role the OECD grouping of countries plays within the Basel Convention. Article VII of the Convention uses the OECD to divide the world into two categories of countries: those who are members of the OECD and those that are not. This categorical divide partly defines which transboundary shipments of hazardous waste in general and e-waste specifically are, or are not, permissible under the Convention. The OECD also publishes many of its own reports and research relevant to the management of e-waste. Many of these publications are freely available online and can be an important resource for e-waste-related research.

reBOOT

reBOOT is a Canadian-based nonprofit organization dedicated to the repair, refurbishment, and reuse of digital devices. It has facilities in Vancouver, Toronto, and Peterborough. reBOOT's programs are divided into three areas of focus: affordable computers for Canadian residents receiving social assistance, providing Canadian registered charities and not-for-profit groups with computing devices such as desktops and laptops, and the creation and maintenance of public Wi-Fi access networks. The organization also provides a destination for individuals and organizations with IT assets they wish to donate to the causes supported by reBOOT. As with similar organizations, reBOOT provides important services that divert used electronic devices away from waste and recycling channels and toward repair, refurbishment, and reuse. In this way, reBOOT's activities support the conservation of materials and energy embodied in devices and mitigates the pollution and waste arising from their manufacturing by extending their useful lives.

Recycled Materials Association

The Recycled Materials Association (ReMA) is the name adopted in April 2024 for what was formerly the Institute of Scrap Recycling Industries (ISRI). The organization is a not-for-profit trade association that advocates on the behalf of its members. The organization currently has over 1,700 members ranging in size from small and medium-sized US-based enterprises to international firms. As the organization's name implies, it covers a wide range of materials for recycling. Major categories of such materials include ferrous and nonferrous metals, glass, textiles, tires and rubber, paper, plastics, and electronics. With respect to electronics, the ReMA brings together scrap materials recyclers but also device refurbishers and information and technology asset disposition (ITAD) companies. ITAD companies generally focus on redeploying business and industrial-scale electronic device assets to new users when previous users upgrade, refresh, or otherwise seek to dispose of their existing IT assets.

The ReMA advocates for legislation and policy at state, federal, and the international levels that will create a favorable commercial environment for the businesses it represents. At the state and federal levels in the United States, the organization advocates for right to repair as well as certified electronics recycling. At the international level, the organization has directly contributed to negotiations at the Basel Convention. ReMA has its own recycling certification program called RIOS (Recycling Industry Operating Standard; see separate profile) that, like other certification programs, is intended to ensure that the recycling of electronics and other materials occur under specified conditions that protect occupational health and safety as well as mitigate environmental impacts.

ReMA is a useful source of data and information for e-waste researchers. The organization publishes periodic snapshots of the international market for recycled materials as well as more specific discussions of individual material categories, such as electronics. A plethora of other materials are available from the organizations website such as policy discussions, advocacy agendas, as well as a curated history of the recycling industry in the United States and the organization's role in it.

Recycling Industry Operating Standard

The Recycling Industry Operating Standard (RIOS) is an independent, third-party standard for recycling operations, including those for e-waste. RIOS was originally developed for the US context in 2003 by the Institute for Scrap Recycling Industries (ISRI, now Recycled Materials Association or ReMA; see separate profile). It was officially released in 2006 and now certifies recycling operations in multiple countries around the world. In 2021, RIOS entered into a business alliance with the R2 certification program that is managed by Sustainable Electronics Recycling International (SERI; see separate profile). The following year, RIOS was accepted by the e-Stewards program (see separate profile) as an alternative environmental management system for recycling certification under that program.

Like other e-waste recycling certification systems RIOS addresses several aspects of recycling, including setting standards for different material streams, occupational health and safety, and environmental impact. Along with e-Stewards and R2, RIOS represents another example of private regulation of the electronics recycling market.

Repair Association

The Repair Association is a trade association based in New York state that advocates for consumer right to repair legislation and a broader policy landscape that supports the interests of independent repair businesses. The association is a major player in organizing advocacy for right to repair legislation at the federal and state levels in the United States. It also works in coalition with advocacy organizations outside the United States that similarly advocate for right to repair. The Repair Association's advocacy extends across multiple industries including agriculture, appliances, automobiles, consumer electronics, information technology more broadly, medical equipment, and resellers. The association points out that electronics have and are becoming embedded in multiple categories of equipment across these industries rather than being confined to consumer electronics such as phones or laptops. Consequently, right to repair is a fundamental issue for keeping technological infrastructure functioning across many sectors of the economy.

The organization's website is an excellent source of up-to-date information on the status of rate to repair legislation in the United States and beyond. Researchers can find an interactive map that links to relevant legislation either passed or under consideration around the world. The website is also an excellent source of data, reports, and policy analysis related to right to repair issues.

Repair Café Movement

The repair café movement is part of a broad civil society movement concerned with reducing waste, including e-waste, and to enhance community connection amongst people who participate. Although the origin of repair cafés is sometimes attributed to an Amsterdam-based community group begun in 2009, the idea of repair cafés is quite widespread and many similar initiatives exist even if they do not explicitly refer to themselves as using the term. Common features of these initiatives include a physical space in which repair activities can take place as well as shared sets of tools and manuals. Repair cafés sometimes specialize in particular device categories, but often involve people fixing a wide range of consumer items—everything including bikes, electronics, textiles, furniture, and other household items.

The physical infrastructure of these cafés provides a place for people to congregate and repair together. Those who do so benefit in several ways. For example, shared sets of tools make repair more affordable as participants do not each have to acquire their own separate tool sets. This kind of approach also means that the environmental impacts of mining for, and manufacturing of, the tools themselves are spread out over a larger number of people rather than an individual. Instead of, say, fifty people each buying fifty specialized sets of tools for electronics repair, those same fifty people can share one or two tool sets between them. This, too, is a form of environmental conservation. Another feature of repair cafés is their support for community building. They provide a space for mixed groups of people to interact, share knowledge, and work together on the common project of repairing things such as electronic devices. In this respect, repair café benefits extend beyond just the conservation of materials and energy embodied in devices to also include the potential to enhance social cohesion.

The Amsterdam-based repair café maintains a website with extensive resources freely available to anyone who can access them online. These resources include a world map that depicts the location of repair café initiatives around the world with links to their webpages, a starter guide for people interested in launching their own repair café, data about repair, links to repair manuals, and repair-related news.

Repair Europe

Repair Europe is a civil society organization advocating for right to repair in the EU and beyond. The organization is composed of over 100 coalition partners across Europe. These partners represent ENGOs, community repair groups, and repair-related businesses of various sizes from sole-proprietor shops to larger device repair chains and spare parts distributors. Repair Europe presents itself as part of a movement rather than a standalone organization.

Repair Europe has been active since 2021 and gained substantial momentum during policy debates over the EU's Green Deal and Circular Economy action plans and changes to legislation such as the Ecodesign Directive and the Directive on Common Rules Promoting the Repair of Goods. The organization focused its campaign on how electronics would be treated under this legislation. After intense lobbying with coalition partners, Repair Europe achieved some success with the directive on repair. The legislation is currently being past among EU member states. It includes rules that require reasonable prices for original replacement parts and a ban on parts pairing (a practice used by manufacturers to thwart the use of third-party components). According to Repair Europe, however, there remains much room for improvement in future iterations of this EU legislation. The organization maintains a continually updated list of resources that include substantive research and commentary on right to repair legislation and associated issues. The work of right to Repair Europe and its coalition partners contributes to an important form of conservation of materials and energy embodied in devices.

Responsible Business Alliance

The Responsible Business Alliance (RBA) is a not-for-profit industry association originally founded in 2004 under the name the Electronic Industry Citizenship Coalition (EICC). RBA's members include major international electronics, retail, auto, and toy companies that voluntarily commit to the RBA's Code of Conduct. The inclusion of automobile and toy companies may at first seem counterintuitive; however, both types of companies are heavy users of electronics. Retailers represent an important part of the overall electronics supply chain that brings finished devices to customers of various kinds. The organization claims that its code of conduct is based on standards under the UN Universal Declaration of Human Rights, the International Labour Organization's labor standards, OECD guidelines for multinational enterprises, and a variety of other established international standards.

The current RBA's Code of Conduct is freely available to download in multiple languages. It covers various criteria categorized under labor conditions, health and safety, environment, ethics, and management systems. The categories for labor conditions, health and safety, and environment address issues of pollution and waste in the electronics sector that can get beyond the narrow framing of e-waste as a post-consumer waste management problem. The organization's Responsible Environment Initiative (REI), for example, focuses on chemical management, decarbonization, circular materials, and water stewardship. The REI chemicals management activities focus on safe use of chemicals, reducing chemical exposures to workers, chemical risk assessments, and finding ways to phase out hazardous chemicals. Decarbonization initiatives seek to reduce or eliminate greenhouse gas emissions from the sector. The REI's circular materials initiative attempts to find ways to incorporate used materials back into electronics manufacturing supply chains. Its water stewardship activities include finding ways to reduce water use as well as minimize or eliminate hazardous chemicals in wastewater discharge from the electronics manufacturing sector. Overall, RBA's Code of Conduct emphasizes pollution prevention and resource conservation as well as the safe handling of hazardous substances. It also pays attention to the need to reduce or eliminate and responsibly dispose of solid waste and emissions to air and water.

Restart Project

The Restart Project began in 2012 as a small community project based in London, UK. Although small in size, its ambitions are to facilitate broad-scale social change regarding peoples' relationships with their electronic devices. Since it began in 2012, the organization has grown substantially and has been instrumental in expanding the network of community-based electronics repair internationally. It cofounded Repair Europe and the Open Repair Alliance (see separate profiles). It also organizes an annual international gathering called Fixfest which brings together activists, policymakers, researchers, educators, and companies to exchange ideas and research on how repair can be better integrated into places where it is not already the norm.

The Restart Project has affinities with the broader repair café movement and similar initiatives. The organization distinguishes itself in part by its focus on electronics and its approach to advocacy which centers on "restart parties." Restart parties are often organized as one off, pop-up-style events held in different venues over time in a given location. They bring together mixed groups of people for fun and intensive shared repairing events usually with volunteers associated with the project who have specialized knowledge about electronics repair. These volunteers work directly with participants partly to demystify electronics repair and share knowledge about it, but also to assist with repairs that are technically challenging.

Although restart parties are a core activity of the organization, the Restart Project supports a range of other electronics repair activities. These initiatives include the analysis of data on repair collected during restart parties and related events. These data are compiled into an online dashboard called the Fixometer that tracks numbers of repair events hosted throughout the networks the organization is part of, the number of participants involved, the number of hours volunteered, and the mass of solid waste avoided, and the CO_2 emissions prevented through electronics repair activities. The Restart Project has also developed free packages of educational material, research on the reuse potential of electronics sent for recycling, a media kit, and the organization produces a monthly podcast.

Santa Clara Center for Occupational Safety and Health

The Santa Clara Center for Occupational Safety and Health (SCCOSH) arose from the work of women's health and labor rights activists concerned about the occupational health and safety conditions of working-class minorities in Silicon Valley in the 1970s. SCCOSH was a leader in pushing for bans on certain chemicals used in electronics manufacturing that were found to be hazardous to human health, especially in terms of women's reproductive health. SCCOSH focused its early activities on toxicants such as trichloroethylene (TCE), two chemicals widely used in various phases of electronics manufacturing. SCCOSH began to document the effects of these and other toxicants on workers, particularly women, in electronics plants in Silicon Valley as well as how these toxicants were contaminating groundwater in the region.

The organization was a leader in advocacy for pollution and waste abatement in electronics manufacturing facilities and their surrounding environments. Indeed, it was the work done by SCCOSH on the environmental effects of toxicants from the electronics manufacturing sector that led to a recognition of a need for broader organizing on these issues beyond just the workplace. In 1982, SCCOSH advocates spawned a new advocacy group called the Silicon

Valley Toxics Coalition (SVTC; see separate profile) to take on the broader environmental impacts of the industry in the region. In these ways, SCCOSH is a forerunner of more contemporary and broader understandings of what counts as e-waste beyond a narrow focus on devices discarded by consumers.

Silicon Valley Toxics Coalition

The Silicon Valley Toxics Coalition (SVTC) was a project of SCCOSH that eventually became a separate entity in 1982. SVTC was originally formed due to the growing awareness of contamination from the electronics industry in Silicon Valley beyond the confines of manufacturing plants and into the broader community, especially into the region's drinking water. A key event was the discovery of leaking chemical storage tanks at a Fairchild Semiconductor plant in 1981. The Fairchild leak was quickly linked with other similar such leaks and other major electronics manufacturing facilities in the valley such as Intel and Raytheon.

SVTC distinguished itself from SCCOSH and other coalition partners through its focus on resisting toxicants from the electronics sector impact in communities and neighborhoods in Silicon Valley. SVTC advocated for changes to county and municipal ordinances that covered how chemicals could be stored, transported, and disposed of in Santa Clara. Among the organization's most important victories were its success in winning a community right-to-know law and a hazardous materials model ordinance. The principle of community right to know has, subsequently, become a core principle of hazardous pollution regulation in the United States and beyond. SVTC's model hazardous materials ordinance required stricter chemical containment infrastructure and monitoring. That ordinance passed in 1983 has become an industry standard since that time. Work done by SVTC and its coalition partners would, among other things, eventually lead the US EPA to designate twenty-nine sites in Santa Clara County eligible for public funds to support remediation (i.e., "Superfund" sites).

Solve the E-waste Problem

Solve the E-waste Problem (StEP) is an international organization created to research policy pathways for the mitigation and elimination of e-waste. It began as an initiative under the United Nations University in 2004 and subsequently evolved into an independent organization based in Vienna, Austria. StEP's work is supported by a tiered system of membership fees. Members include representatives of large, transnational device manufacturers and software companies, government ministries, trade associations, ENGOs, reverse logistics and recycling firms, universities and research institutes, and individual researchers.

StEP focuses its work on four categories of action: research, strategy and goalsetting, training and development, and communication and awareness raising. The organization efforts in these areas include publishing publicly available reports and position papers on a wide range of e-waste-related issues. For example, StEP's early research focused on designing collection systems for e-waste and policy to support them. It also did early work on attempts to provide universal definitions of e-waste for international regulation. Some of the latter work fed into specifications of technical methods for estimating global comparable statistics about e-waste arising at the national level in countries around the world. Related research initiated by StEP led to the publication of a publicly available online map of e-waste facts, figures, and regulations around the world. That Initiative later branched off into a separate organization called the Global E-waste Partnership (see separate profile). More recent research and policy work by StEP includes suggestions for linking formal and informal recycling sectors under EPR policies and projections of future e-waste generation under a variety of alternative economic scenarios. In terms of training and capacity building, StEP organizes "summer schools" that bring college and university students together with researchers, policymakers, and representatives of device manufacturers and recyclers for intensive week-long experiential learning programs.

Sustainable Electronics Recycling International

Sustainable Electronics Recycling International (SERI) is a nonprofit organization dedicated to the reuse and recycling of electronics in environmentally safe ways that also protect the health and safety of workers and communities. SERI explicitly notes that reuse and recycling are necessary, but insufficient on their own to reduce and eliminate the pollution and waste arising from electronics. The organization recognizes that a broader understanding of e-waste that

162 *Electronic Waste*

includes pollution and waste arising before device users discard them requires changes in how electronics are designed and manufactured as well.

A primary activity of SERI is operating the R2 recycling certification standards. R2 is one of two US-based electronics certification standards, the other being e-Stewards (see separate profile). R2 was initially championed under the auspices of the US EPA, beginning in 2006. Differences of opinion between stakeholders in that process led to a split in which BAN (see profile) developed e-Stewards and ISRI (now ReMA) supported R2. Eventually, R2 came to be housed under SERI, a standalone organization. As of 2024, over 1,000 facilities in forty-one countries have attained R2 certification. Like e-Stewards the R2 System relies on a network of not-for-profit standardization bodies and for-profit auditing companies to certify individual facilities and recycling companies. In this respect, R2 shares commonalties with other electronics recycling certification systems of being a form of private regulation introduced to shape the behavior of market participants.

United Nations Conference on Trade and Development

The United Nations Conference on Trade and Development (UNCTAD) is an intergovernmental body housed within the UN Secretariat. Its main goal is to facilitate economic integration and cooperation between countries. The details of the organization's mission have changed since the body formed in 1964, but it currently includes increasing access to digital technologies among its core efforts. As the name of the organization implies, it is concerned with a very broad mandate around all aspects of the global economy. However, UNCTAD produces its own projects and reports on e-waste management as well as being a partner on multiple related projects of other organizations.

E-waste researchers will find the UNCTAD website a useful source for high-level intergovernmental reports and documents related to the topic. For example, in the organization's *Data for Development* report released in May 2024 researchers can find a section of the report devoted to the politics of e-waste. That the topic is included in a report otherwise devoted to broad issues of economic development suggests the importance of e-waste as a matter of concern for high-level policymakers. Beyond its own reports and documents, UNCTAD's website is a useful source of links to other organizations with relevant information and data about e-waste.

United Nations Environment Programme

The United Nations Environment Programme (UNEP) was established in 1972 and is headquartered in Nairobi, Kenya. It is the primary UN body for coordinating responses to environmental issues within the UN and its member states. UNEP works directly on the issue of e-waste within its broader remit. Some of this work approaches e-waste from the perspective of managing post-consumer discards, but the work of UNEP also extends to broader understandings of e-waste via the organization's work on pollution and waste arising from mining and manufacturing. For example, UNEP houses the secretariats of three major international conventions that impact the manufacturing of electronics and the management of post-consumer e-waste: the Rotterdam Convention, the Stockholm Convention, and the Basel Convention. Rotterdam and Stockholm focus on reducing and eliminating various kinds of chemical hazards and pollution upstream in manufacturing. They cover all industries in which these chemicals are in use, including the electronics manufacturing sector. The Basel Convention addresses the international trade in traffic of post-consumer hazardous waste, including e-waste.

UNEP is a vital source of information and data for e-waste researchers. More than 100 reports on e-waste can be accessed from UNEP's website. The organization's "World Environment Situation Room" also hosts data and numerous reports on related issues. For example, researchers can access data and information on chemicals and waste that are relevant to the manufacturing of electronics. This makes UNEP an important source of data and information about the broader meaning of pollution and waste arising throughout the lifecycle of electronics.

United Nations Industrial Development Organization

The United Nations Industrial Development Organization (UNIDO) was established by the UN General Assembly in 1966 and is headquartered in Vienna, Austria. In its original incarnation the organization focused on assisting the economic transformation of so-called developing countries. Over time, the organization's mandate has broadened to the facilitation of industrial development writ large, although it still maintains a developmentalist orientation toward the international economy. Within its broader mandate, UNIDO continues to play an important role in international research and policymaking about e-waste. The organization sponsored much of the early country- and region-specific research on e-waste, particularly in Africa. UNIDO continues to sponsor training, research, and policy analysis on e-waste management. Today these initiatives are framed within the UN's seventeen sustainable development goals articulated under the UN General Assembly's *2030 Agenda for Sustainable Development*. The organization's website is an important source of material for e-waste researchers. Numerous UNIDO reports, news items, policy analysis, and related documents are available for free access.

United Nations Institute for Training and Research

The United Nations Institute for Training and Research (UNITAR) is a capacity-building, training, and research organization within the broader UN system. It was established in 1963 and is headquartered in Geneva, Switzerland. Like other UN organizations, UNITAR has a broad mandate and, as part of that, it addresses specific research and training issues around e-waste. For example, UNITAR is an important sponsor of the Global E-waste Statistics Partnership. The organization also sponsors country and region-specific policy analysis and research on e-waste, particularly in so-called developing countries and regions in Africa and Asia.

In 2022, UNITAR launched its "Sustainable Cycles" program with the overall goal of promoting sustainable societies. The initiative has a particular focus on sustainable production, consumption, and disposal of electrical and electronic equipment. Work under the program is categorized into four foci: quantification, waste management systems, partnership initiatives, and training and education. A plethora of data and research related to a broad understanding of pollution and waste arising from electronics, from manufacturing to disposal, is available to researchers at UNITAR's Sustainable Cycles website.

United Nations University

The United Nations University (UNU) was established in 1972 with headquarters in Tokyo, Japan, and several satellite campuses around the world. UNU is both a research university and think tank under the broader United Nations system. Like other UN initiatives, the university has a broad remit to research and solve global issues associated with human development and welfare. UNU-based researchers published some of the earliest research on the environmental impacts of electronic devices. Importantly, this early research took a broad understanding of the environmental impact of digital technologies that included those related to manufacturing, not just post-consumer discards. In this sense, this research is an important precursor to the broader understanding of e-waste. UNU was an important founding partner of the Solve the E-waste Program (StEP) initiative. Researchers at UNU are important contributors to the Global E-waste Statistics Partnership and other e-waste initiatives associated with UNITAR's Sustainable Cycles program.

United States Federal Trade Commission

The United States Federal Trade Commission (FTC) is a US government agency whose principal mission is the enforcement of antitrust law and consumer protection. The agency formed in 1914 largely in response to the crises of monopoly business formation, particularly those associated with oil, tobacco, and railway interests. In the early 2010s, the FTC began to intervene in the growing problem of corporate monopolization of repair and the parallel erosion of consumers' right to repair. Some of this early work related to these issues in the auto industry, but soon expanded into the realm of consumer electronics.

A key event organized by the FTC in 2019 enhanced public advocacy for consumers' right to repair. The event was called *Nixing the Fix* and saw a public testimony from citizens, consumers, independent repair business proprietors, and major brand manufacturers. Among the findings from that public hearing was that many brand manufacturers were engaging in illegal behavior by using deceptive warning labels on and in their devices that suggested unauthorized repairs would result in voided warranties. Since the *Nixing the Fix* event the FTC has made right to repair a policy priority. The agency released a major report and policy statement following the event. It also pursues ongoing regulatory action and legal cases against companies violating consumer right to repair. In particular, the FTC argues that companies are restricting consumers' right to repair in ways that violate the 1975 Magnuson-Moss Warranty Act. Among other things, this act prohibits business arrangements that put conditions on consumer product's warranties as they relate to independent third-party service

166 *Electronic Waste*

providers such as those for repair. The regulatory action of the FTC on right to repair is a crucial intervention that can reduce the generation of post-consumer e-waste. E-waste researchers interested in right to repair will find the FTC website to be an excellent source of policy discussion, legal cases, and related information.

United States Public Interest Research Group

The United States Public Interest Research Group (US PIRG) is a nonprofit civil society advocacy organization, the beginnings of which can be traced to student activism on US college campuses in the 1970s. US PIRG incorporated in 1984 and is headquartered in Denver, Colorado. The organization is focused on a wide range of issues related to consumer protection, public health, and transportation. Starting in the early 2010s, US PIRG joined other US-based groups advocating for right to repair with a particular focus on electronic technologies.

US PIRG continues to play an important role in advocating for consumer right to repair legislation in the United States. Representatives of the group have directly participated in lobbying state legislatures and US Congress. The group has also supported research on consumer support for right to repair, independent repair businesses experiences with growing challenges to device repair, and other related research. US PIRG's website is a useful source for e-waste researchers. Research and reports are freely available as are frequent News updates about the state of right to repair. Although the work is mostly US-focused, some information about international efforts around right to repair is also available.

WEEE Forum

The Waste Electrical and Electronic Equipment Forum (WEEE Forum) is a nonprofit consortium of producer responsibility organizations (PROs) founded in 2002 and headquartered in Brussels, Belgium. The organization's current membership includes over fifty PROs, mostly in the EU and Scandinavia, but also those based in Australia, Canada, Colombia, India, New Zealand, Nigeria, South Africa, and the United States. WEEE Forum's main goal is to advocate for a harmonized system of electronic waste collection, logistics, and processing. To that end it maintains activity on several campaign fronts. These include circular economy, collection of WEEE, critical raw materials, data collection, illegal trade, and standards.

PROs have become the dominant governance mechanism for e-waste take back systems in many jurisdictions around the world. Typically, PROs are comprised of representatives from electronics brand manufacturers and retailers, though there can be differences of detail between jurisdictions. These nonprofit NGOS are granted responsibility for collecting environmental handling fees charged to consumers when they purchase new electronics. The PROs then disperse those funds to collection and processing businesses that operate in a given jurisdiction. The WEEE Forum website provides a plethora of links to data, policy analysis, and reports as well as a resources page that researchers can consult for access to a wide range of issues related to e-waste from the perspective of PROs.

WEEELABEX

WEEELABEX is a not-for-profit organization managed by electrical and electronic equipment manufacturers to support a third-party certification system for WEEE collection and processing in Europe. WEEELABEX began as a research project co-funded by the EU and the WEEE Forum. The project began in 2009 with the goal of coordinating the multiple WEEE collection systems that existed in different EU jurisdictions into a harmonized system for the region as a whole. During the project lifetime from 2009 to 2011 up to twenty-five different WEEE compliance schemes existed in the EU. The consortium of stakeholders that would eventually become the WEEELABEX organization publicly released its standards document in 2011. WEEELABEX became a standalone nonprofit entity in 2013. Since then, it has housed the proprietary auditing standards that certify processors who seek WEEELABEX certification.

As is the case with other third-party certifications such as e-Stewards, RIOS, and R2, WEEELABEX is a form of private regulation that has been folded into the legal frameworks of state jurisdictions. Similarly, WEEELABEX relies on a network of private third-party auditing companies to perform its proprietary certification process. The WEEELABEX website maintains an updated list of certified e-waste processors that includes information about their location and the materials they do and do not handle, among other information. This kind of information can be a useful source of data for e-waste researchers interested in understanding the landscape of certified electronic waste processing in the EU region.

World Computer Exchange

The World Computer Exchange (WCE) is a US-based nonprofit organization founded in 1999. WCE focuses on computer refurbishment and repair for donation to communities in what the organization calls developing countries. WCE maintains chapters in multiple US states, Puerto Rico, and several African countries including Liberia, Kenya, and Zambia. The organization supports multiple campaigns that boost the useful lifetimes of computers. This activity acts as a form of material and energy conservation. WCE's initiatives also include supporting electronics refurbishing clubs, coding boot camps, and digital literacy programs. The organization's work to share technical skills for hardware refurbishment and repair are an important intervention into the mitigation of post-consumer e-waste. They also provide an entry point for people who may move on from volunteer skills shares to operating their own independent repair and refurbishment organizations, including for-profit micro-enterprises.

World Health Organization

The World Health Organization (WHO) is the United Nations agency that specializes in public health. It was formed in 1948 and is headquartered in Geneva, Switzerland. As might be expected, WHO's remit is broad: it is tasked with promoting universal health care across the world with a particular emphasis on those most vulnerable to adverse health situations. WHO began taking a particular interest in the health impacts of electronic waste processing, especially as they relate to children, beginning in approximately 2018. Electronic waste was incorporated into the organization's global health strategy articulated in 2019. Since then, WHO has developed several e-waste public health initiatives with a particular emphasis on women and child laborers in the informal recycling sector. In this respect, the WHO approach to e-waste treats the issue as a concern of post-consumer discards.

WHO's concerns about e-waste focus on exposure via scavenging, the dumping of scrap electronics on land or in water bodies, the use of open burning and acid baths or leaching for materials recovery, as well as manual stripping, shredding, and this assembly of equipment. To solve these problems, WHO recommends, among other things, the enforcement of existing international agreements such as the Basel Convention, the development and enforcement of legislation for the environmentally sound management of e-waste, the implementation of appropriate occupational health and safety measures for people who work in the e-waste processing sector, and the elimination of child labor. The WHO website offers a plethora of resources freely available to e-waste researchers. These resources include short documents on key facts as well as substantive reports on e-waste and public health. The website also offers a massively open online course (MOOC) on e-waste that is freely available to the public.

World Reuse, Repair, and Recycling Association

The World Reuse, Repair, and Recycling Association (WR3A) is a not-for-profit organization established in 2006 and based in the US state of Vermont. The organization advocates for a fair-trade model for e-waste collection, repair, and recycling using existing fair trade models for other commodities, such as coffee, for inspiration. Toward this end, the organization has advocated against bans on exports of used and scrap electronics from the United States and other jurisdictions. WR3A offers an alternative model of environmentally sound management of e-waste that differs from international agreements like the Basel Convention which emphasizes bans on exports.

WR3A's fair trade model involves relying on the legal enforceability of contracts between exporters and importers of used, scrap, and waste electronic equipment. The organization advocates for the use of such contracts to incentivize the proper collection and processing of electronic equipment that is beyond use. Depending on the specific trade arrangement in question, this may mean that an importer based in a country with insufficient e-waste recycling infrastructure may collect locally available e-waste for export to a jurisdiction with the appropriate recycling infrastructure. In exchange, the importer of used but serviceable and repairable electronics receives a price premium for the equipment locally collected and lower prices for the used but serviceable and repairable equipment to be imported. The net result is that used electronics

170 *Electronic Waste*

are moved to markets where they can and will be put to use for longer periods of time, thus enabling the conservation of materials and energy embodied in those devices. Meanwhile, electronic equipment no longer able to be repaired is properly collected and moved to markets where appropriate electronic waste recycling and disposal infrastructure exist.

WorldLoop

WorldLoop is a Europe-based not-for-profit NGO dedicated to eliminating the harmful environmental impacts of post-consumer e-waste processing in what it calls developing countries. The organization got its start in 2008 as a project of a social enterprise called Close the Gap. Close the Gap sought to deal with international inequities of access to digital technology by sending donated, but reusable equipment to countries where access rates to IT equipment are low. WorldLoop began as a pilot project dedicated to collecting and properly handling post-consumer waste from the jurisdictions receiving donated used equipment from Close the Gap. The success of the pilot project led to the spinoff of WorldLoop as a separate standalone organization.

To achieve its goals, WorldLoop uses a two-prong approach. Its parent organization, Closing the Gap, coordinates donations of reusable electronic equipment to recipient organizations and projects, mostly in countries of Africa. The second part of WorldLoop's approach is to coordinate environmentally sound collection of post-consumer e-waste from the same locations that receive donated equipment from Closing the Gap. Equipment so collected is then processed in facilities set up for the proper handling of waste electronics. This may mean that such equipment is locally collected and then exported to jurisdictions with appropriate infrastructure or, depending on the circumstances, WorldLoop may facilitate the set-up of appropriate e-waste recycling infrastructure in situ. WorldLoop offers a model of e-waste handling like WR3A's in this respect.

WorldLoop's annual reports can be an important source of information for e-waste researchers. These reports detail the organization's activities and ongoing support of associated projects that extend beyond its original focus on supplying reusable equipment and responsibly collecting post-consumer e-waste. Such projects include various kinds of IT-enabled makerspaces, circular economy and innovation hubs, and similar social enterprises. The organization also publishes data about the environmental offset gained through its e-waste recycling activities.

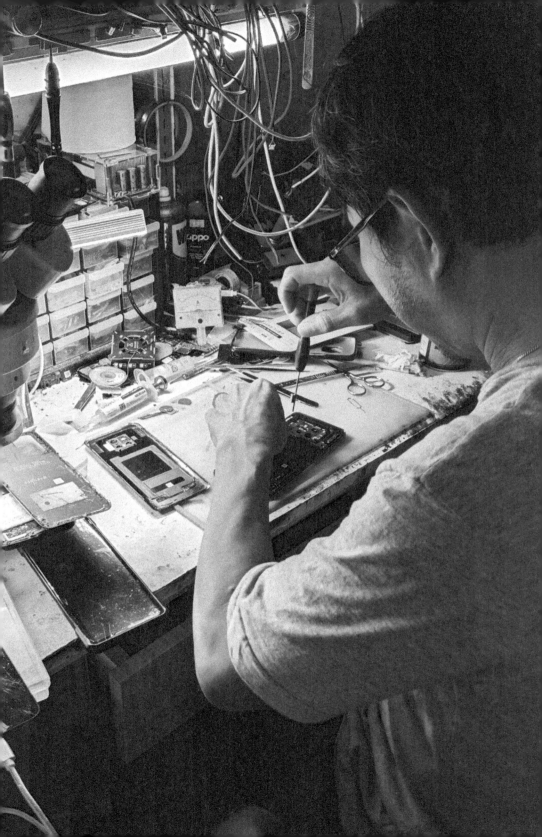

5

Documents

This chapter presents primary resources for further research into the topic of pollution and waste arising from electronics. It presents and contextualizes excerpts from a variety of primary documents that describe pollution and waste issues relating to mining, manufacturing, post-consumer discarding, and repair of electronics.

"Recovery of Precious Metals from Electronic Scrap," United States Department of the Interior (1972)

In 1972, the United States Department of the Interior under the Bureau of Mines published a study of practical methods for extracting precious metals from electronic scrap. The resulting report is one of the earliest publications dealing with discarded electronics. Among other things, what is notable about the report is that there is no reference at all to discarded electronics as "waste." In this respect, the report is a good demonstration of the fluid relationship between "waste" and "resource."

Electronic scrap represents a significant source of secondary precious metals [...] Ultimately, most of this electronic equipment becomes obsolete or damaged and is scrapped. Some of the scrapped items may be salvaged and reused as electronic components. The remainder represents a potential source of secondary precious and base metals. Much of the higher grade scrap is processed for precious metal

---

A technician repairs a phone in his workshop in Perak, Malaysia. Repair and reuse are important strategies for reducing e-waste, and the last decade or so has seen increased advocacy for right to repair legislation around the world. Some manufacturers have lobbied against this type of legislation or attempted to flout its enforcement through various means. (Swee Ming Young/Dreamstime.com)

174   *Electronic Waste*

recovery; however, at present significant quantities of precious metals are being lost in scrap which is too low in precious metal content or too complex to warrant recovery. The Bureau of Mines undertook the development of a practical process to recover precious metals from complex low-grade scrap which presently is not being treated [...]

From a metallurgical standpoint, electronic scrap is a complex mixture of various metals attached to, covered with, or mixed with diverse types of plastics and ceramics. Precious metals occur as platings of various thicknesses, relay contact points, switch contacts, and wires, or in solders. As electronic circuits become more sophisticated, the proportion of precious metals in relation to other metals increases [...]

Source: Dannenberg, R. O., J. M. Maurice, and G. M. Potter. 1972. "Recovery of Precious Metal from Electronic Scrap." Salt Lake City, UT: United States Department of the Interior, Bureau of Mines.

"Flows of Selected Materials Associated with World Copper Smelting," United States Geological Survey (2005)

The electronics industry is the second largest consumer of copper after the building and construction industry. The United States Geological Survey publishes detailed reports and data on copper and other important minerals. This report provides a detailed analysis of the ecological footprint of copper extraction from major mining operations around the world. Data from this report on the Chuquicamata copper mine is highlighted in Chapter 2. What is notable in the following excerpt from the report is its finding that 1 kg of economically useful copper results in 210 kg of mine waste and additional waste by-products.

The United States Geological Survey collects, analyzes, and disseminates information about the material flows associated with mineral extraction and use. This publication focuses on the flows associated with the smelting of copper concentrates. Copper is one of the core commodities in modern and developing economies. One cost paid for placing it in service is the significant, though manageable, set of environmental effects associated with the smelting process.

Through a detailed assessment of smelting technologies, covering approximately two-thirds of world capacity, this research quantifies the unwanted by-products associated with copper production. The conversion of

Documents 175

1 kilogram (kg) of copper concentrate from its in-ground condition (ore) into economic service generates an average landscape footprint comprised of 210 kg of mine waste, 113 kg of mill tailings, 2 kg of slag, and 2.3 kg of sulfur-bearing co-product. The corresponding air releases per kg of copper include 0.5 kg of carbon dioxide and 0.2 kg of sulfur dioxide.

Because copper concentrates are a traded worldwide, the ecological footprint described above can be allocated among the countries that smelt and export copper and to the countries that import and use it. This report also highlights the relationship between exporting and importing countries.

Source: United States Geological Survey, and Thomas G. Goonan. 2005. "Flows of Selected Materials Associated with World Copper Smelting." Reston, VA: United States Geological Survey. https://pubs.usgs.gov/of/2004/1395/.

"Fairchild, Intel, and Raytheon Sites Middlefield/Ellis/ Whisman (MEW) Study Area Mountain View, California, Record of Decision," United States Environmental Protection Agency (1989)

Fairchild Semiconductors, Intel, and Raytheon are three of the most important early electronics manufacturers in the region sometimes called Silicon Valley. Both Fairchild and Raytheon have been absorbed into other corporations, but Intel remains a corporate entity. This report from the United States Environmental Protection Agency describes some of the publicly available data that went into designating the locations around the original Fairchild, Intel, and Raytheon facilities as being part of a number of Superfund sites in the area. The findings described in the report are fairly technical; however, they are indicative of a broader and more important issue which is that electronics manufacturing is an important source of pollution and waste released to the environment. This report makes it clear that despite the marketing images of electronics being "clean" and "light" industries, making these devices entails significant contamination of surrounding environments.

Site Location and Description

The Middlefield/Ellis/Whisman (MEW) Study Area is located in Santa Clara County in the city of Mountain View, California. The site is divided into a Local Study Area (LSA) and a Regional Study Area (RSA). [...] The LSA consists of three National Priority List (NPL) sites (Fairchild, Intel and Raytheon), as well as several non-Superfund sites. The LSA encompasses about 1/2 square mile of the RSA and contains primarily light industrial and commercial areas, with some residential areas west of Whisman Road. The RSA encompasses approximately 8 square miles and includes Moffett Naval Air Station (an NPL site) and NASA Ames Research Center, along with light industrial, commercial, agricultural, park, golf course, undeveloped land, residential, motel and school land uses.

Various owners or occupants in the area around the intersections of Middlefield Road, Ellis Street, Whisman Road, and the Bayshore Freeway (US Highway 101), are or were involved in the manufacture of semiconductors, metal finishing operations, parts cleaning, aircraft maintenance, and other activities requiring the use of a variety of chemicals [...] Site investigations at several of these facilities have revealed the presence of toxic chemicals in the subsurface soils and groundwater. To investigate the extent of groundwater contamination emanating from the LSA, and soil contamination at their respective facilities, Fairchild, Intel, and Raytheon performed a Remedial Investigation and a Feasibility Study of potential remedial alternatives under the direction of EPA.

There are no natural surface drainage features within the Local Study Area. The nearest significant natural surface drainage features of the Regional Study Area are Stevens Creek to the west and Calabazas Creek to the east. Calabazas Creek is located approximately four miles east of the MEW Study Area. Stevens Creek forms the western boundary of the Regional Study Area. Both discharge into the San Francisco Bay. Surface water runoff from most of the RSA and all of the LSA south of the Bayshore Freeway is intercepted by a storm drain system and is discharged into Stevens Creek. To the north of the Bayshore Freeway, most of the runoff from Moffett Field Naval Air Station is collected by a storm drain system that ultimately discharges to Guadalupe Slough of San Francisco Bay. Runoff from the northwestern portion of Moffett Field discharges into Stevens Creek.

The Local and Regional Study Areas are underlain by a thick sequence of unconsolidated sediments deposited into a structural depression. The sediments are comprised of alluvial fan, estuarine, and bay mud deposits. Repeated variations in sea levels resulted in a complex sedimentary sequence characterized by irregular interbedding and interfingering of coarse and fine grained deposits.

Groundwater aquifers at the site are subdivided into shallow and deep aquifer systems, separated by a laterally extensive regional aquitard. The shallow aquifer system comprises aquifers and aquitards to a depth of approximately 160 feet below the surface. Within the shallow system four primary hydrogeologic aquiferzones have been identified based upon the occurrence of aquifer material and a similar depth below the surface. The shallow aquifer system is comprised of the A-aquifer and the underlying B1-, B2-, and B3-aquifers. The regional B-C aquitard separates the B3-aquifers from the C-aquifer and the deep aquifer system. Current groundwater flow in aquifer zones above the B-C aquitard is generally to the north, toward San Francisco Bay.

Site History

During 1981 and 1982, preliminary investigations of facilities within the LSA indicated significant concentrations of contaminants in soil and groundwater. By 1984, the Fairchild, Intel and Raytheon sites, located within the LSA, were proposed for the Federal National Priorities List (NPL). By 1985, five companies within the LSA (Fairchild, Intel, Raytheon, NEC, and Siltec) initiated a joint investigation to document and characterize the distribution of chemicals emanating from their facilities. In April 1985, the California Regional Water Quality Control Board—San Francisco Bay Region (RWQCB) adopted Waste Discharge Requirements (WDRs) for each of the five companies. The primary cause of the subsurface contamination was from leaking storage tanks and lines, and poor waste management practices.

On August 15, 1985, Fairchild, Intel, and Raytheon entered into a Consent Order with the EPA, the RWQCB, and the California Department of Health Services (DHS). Since signing of the Consent Order, the three companies have carried out an extensive Remedial Investigation and Feasibility Study (RI/FS) of chemicals emanating from the LSA and soil contamination at their respective facilities. Work has been performed under the supervision of EPA, the RWQCB, DHS, and the Santa Clara Valley Water District (SCVWD). Prior to and during the site investigation, the companies have been conducting interim clean up activities at the site. These interim remedial actions include tank removals, soil removal and treatment, well sealing, construction of slurry walls, and hydraulic control and treatment of local groundwater. NEC and Siltec declined to enter into the Consent Order and were placed under RWQCB enforcement authority [...]

Nature and Extent of Contamination

Industrial activities conducted within the MEW Study Area required the storage, handling and use of a large number of chemicals, particularly solvents and other chemicals used in a variety of manufacturing processes. Significant quantities of volatile organic chemicals were used for degreasing, process operations, and for general maintenance. Raw and waste solvents and other chemicals were piped and stored in underground systems. The presence of chemicals in the subsurface soils and groundwater, that originated from facilities in the MEW area, are primarily the result of leaks from these subsurface tanks and lines, sumps, chemical handling and storage areas, and utility corridors. Chemical releases occurred, for the most part, below the ground surface and migrated downward into the aquifer system.

Investigations at the site have revealed the presence of over 70 compounds in groundwater, surface water, sediments, and subsurface soils. The vast majority and quantity of these compounds are found in groundwater and subsurface soils. Three major classes of chemicals were investigated [...] (1) volatile organic compounds, (2) semi-volatile acid and base/neutral extractable organic compounds, and (3) priority pollutant metals. Of these three classes, volatile organics are the most prevalent.

An extensive area of groundwater contamination has been defined [...] Current site data indicate that chemicals are present primarily in the A-, B1-, and B2-aquifer zones. To a much lesser degree, chemicals have been detected in localized areas of the B3-,C-aquifer, and deeper aquifer zones. Contamination of the C-aquifer and deeper aquifers appears to have resulted from chemicals migrating downward from shallow areas containing elevated chemical concentrations, through conduit wells, into groundwater of the deep aquifer system. The C and Deep aquifers most affected by contamination are in the areas of the so-called Rezendes Wells, located near Fairchild Building 20, and the Silva Well, located at 42 Sherland Avenue. These wells have subsequently been sealed. The closest municipal water supply well, Mountain View #18 (MV 18), is located approximately 1800 feet to the southwest of the Silva Well. Groundwater samples are collected from MV 18 on a regular basis. No contaminants have been identified in any water samples from MV 18. As part of the Remedial Design and Remedial Action (RD/RA) some additional groundwater investigations may be necessary, particularly in the Silva Well area.

Subsurface soil contamination has been found at the Fairchild, Intel, and Raytheon facilities, along with the facilities of other PRPs within the RSA. Trichloroethene (TCE), 1,1,1trichloroethane (TCA), trichlorotrifluoroethane (Freon-113), 1,1dichloroethene (1,1-DCE), 1,2-dichloroethene (1,2-DCE), methylene chloride, toluene, acetone, and xylene are the chemicals most commonly detected in subsurface soils in the LSA. Chemicals associated with activities in the RSA appear to be concentrated in shallow soils above approximately 50 feet or roughly extending to the B1-aquifer. Chemicals are not found in surface soil samples (upper one foot of soil) and do not appear in soils and clay of the B-C aquitard. Chemical found in subsurface soil samples are generally similar to those found in adjacent groundwater samples. As part of the Remedial Design and Remedial Action some additional soil investigations may be necessary in certain areas.

Source: United States Environmental Protection Agency. 1989. "Fairchild, Intel, and Raytheon Sites Middlefield/Ellis/Whisman (MEW) Study Area Mountain View, California | Record of Decision." https://semspub.epa.gov/work/09/88167245.pdf.

"2011–2017 Greenhouse Gas Reporting Program Industrial Profile: Electronics Manufacturing Sector," United States Environmental Protection Agency (2018)

The United States Environmental Protection Agency periodically publishes reports on greenhouse gas emissions from specific economic sectors. This report is focused on the electronics manufacturing sector in the United States. It is an important example of upstream pollution and waste arising from the lives of electronics.

Highlights

Emissions reported by this sector decreased by one-and-a-half percent from 6.2 million metric tons of carbon dioxide equivalent (MMT CO_2e) in 2016 to 6.1 MMT CO_2e in 2017.

This decrease is associated with a decrease in the emissions of sulfur hexafluoride (SF6) and other fully fluorinated compounds.

About This Sector

The Electronics Manufacturing sector includes, but is not limited to, facilities that manufacture semiconductors (including light-emitting diodes), micro-electromechanical systems (MEMS), liquid crystal displays (LCDs), and photovoltaic (PV) cells. Specifically, this sector consists of electronics manufacturing facilities with production processes that use plasma-generated fluorine atoms and other reactive fluorine-containing fragments to etch thin films, clean chambers for depositing thin films, clean wafers, or remove residual material. The sector also includes electronics manufacturing facilities with chemical vapor deposition processes or other production processes that use nitrous oxide (N_2O), and processes that use fluorinated greenhouse gases (GHGs) as heat transfer fluids (HTFs) to control temperature or clean surfaces [...]

From 1995 through 2010, a segment of the U.S. semiconductor manufacturing industry reported their aggregate emissions from the use of fluorinated greenhouse gases (F-GHGs) to the U.S. Environmental Protection Agency (EPA) under the voluntary Perfluorocarbon (PFC) Reduction/Climate Partnership. These manufacturers, representing about 80% of U.S. semiconductor manufacturing capacity, significantly reduced their emissions from etching and chamber cleaning between 1995 and 2010 [...] even though overall production increased during this period. The methods used by facilities to monitor their emissions under the partnership are believed to have been roughly comparable to those used to date [...]

Reported emissions from the Electronics Manufacturing sector decreased from 7 MMT CO_2e in 2011 to 5.2 MMT CO_2e in 2013, a decrease of 25.7 percent. This decrease was primarily due to a large reduction in combustion emissions at one facility. Emissions increased by 19.2 percent between 2013 and 2014, due in part to a rule revision that changed the emission factors used by facilities to estimate emissions, resulting in higher estimated emissions. Emissions decreased from 2014 (6.2 MMT CO_2e) to 2017 (6.1 MMT CO_2e), a reduction of about 0.1 MMT CO_2e (about one-and-a-half percent). The decrease was not due to reductions at any specific facilities, but instead a general decrease across the industry. Most of the reductions were from decreased emissions of PFCs and other fully fluorinated GHGs [...]

Opportunities for Emissions Reductions

Emissions from the Electronics Manufacturing sector can be reduced through a variety of measures that target F-GHG, N_2O, and HTF emissions. To target F-GHG emissions, mitigation options currently used include the NF3 remote chamber cleaning process, gas replacement, process optimization, and installation and use of abatement systems.

As compared to in-situ chamber cleaning processes, the remote cleaning process utilizes a larger portion of the F-GHG being used to clean chemical vapor deposition chambers, resulting in less unreacted gas being emitted. For gas replacement, some F-GHGs used in particular processes may be replaced with more efficient and/or lower GWP gases. Process optimization involves reengineering a process to more efficiently use F-GHGs. Both gas replacement and process optimization are generally used provided that these changes do not negatively impact the production yield.

Various types of abatement are available to mitigate F-GHG emissions from the Electronics Manufacturing sector. These include thermal abatement, catalytic abatement, or plasma abatement. Typically, these are point-of-use abatement systems; however, recent developments through Clean Development Mechanism projects in Asia have shown that centralized abatement has worked for reducing emissions from flat panel display manufacturing. In addition to being used on new facilities, abatement systems for F-GHGs can be retrofitted on existing facilities as well. Abatement systems also are available for N2O emissions. Information submitted through the GHGRP indicates that approximately 35% of U.S. electronics facilities that reported to the GHGRP are using F-GHG and N_2O abatement, indicating there are further opportunities for the use of abatement to reduce Electronics Manufacturing sector emissions.

HTF emissions occur mainly from leakage. To reduce HTF emissions, proper handling and equipment maintenance techniques can be implemented to mitigate equipment leaks.

According to the International SEMATECH Manufacturing Initiative as of 2005 the semiconductor manufacturing industry was taking the following steps to reduce PFC emissions:

- Decommissioning fabrication plants manufacturing 150 millimeter or smaller wafers,
- Installation of abatement equipment,
- Process optimization,
- Installation of endpoint detection for processes to minimize gas consumption,
- Use of new and alternative clean chemistries,

182   *Electronic Waste*

- Integration of low emissions chemical vapor deposition (CVD) tools, and
- Increasing wafer size and advanced process technology.

Additional opportunities for emissions reductions include PFC replacement and capture and recovery before emissions are released to the atmosphere. (United States Environmental Protection Agency 2018, 13)

Source: United States Environmental Protection Agency. 2018. "2011–2017 Greenhouse Gas Reporting Program Industrial Profile: Electronics Manufacturing Sector." https://www.epa.gov/sites/production/files/2018-10/documents/electronics_manufacturing_2017_industrial_profile.pdf.

"Center for Corporate Climate Leadership Sector Spotlight: Electronics," United States Environmental Protection Agency (2016)

The United States Environmental Protection Agency operated a website hub called the Center for Corporate Climate Leadership. The information and tools at that website were made available for business organizations seeking to measure and manage greenhouse gas emissions from their operations. Earlier versions of the hub hosted webpages and reports on specific sectors of the economy, including the electronics sector. The excerpts here are from a webpage at the hub devoted to fluorinated greenhouse gas emissions (F-GHGs) from the electronics sector, specifically from flat panel display manufacturing. Among the important information in the excerpts is a concrete instance of the rebound effect or Jevons Paradox. The webpage cites reports that although manufacturers in the sector have made substantial reductions in F-GHG emissions the companies have fallen short of their goals because the total production of flat panel displays has increased over time by both by their own facilities and newer competitors entering the market.

Fluorinated greenhouse gases (F-GHGs) are among the most potent and persistent greenhouse gases (GHGs) contributing to global climate change. These gases play a vital role in the manufacture of flat panel displays, namely liquid crystal display (LCD) panels that go into televisions, computer monitors, and many other display products. The overall climate impact of the millions of display products Americans use can be greatly reduced if suppliers of these components take steps to mitigate releases of these F-GHGs to the atmosphere.

Starting in late 2013, brands and retailers Wal-Mart, Dell, HP, Lenovo, Best Buy and Acer took an important step to encourage further F-GHG reductions among their LCD suppliers by asking suppliers to develop a standard method for measuring F-GHG emissions, set new voluntary F-GHG emissions reduction goals with public timelines for demonstrating progress, and develop annual progress reports [...]

What Are F-GHGs and How Are They Used in Panel Manufacturing?

Fluorinated GHGs such as certain perfluorocarbons (e.g., CF_4, C_2F_6, C_4F_8), trifluoromethane (CHF_3), nitrogen trifluoride (NF_3), and sulfur hexafluoride (SF_6) are among the most potent greenhouse gases (GHGs), with some persisting in the atmosphere for thousands of years effectively causing irreversible impacts on the earth's climate system. These gases are commonly used in many types of electronics manufacturing, including the manufacture of flat panel displays, semiconductors, micro-electro-mechanical systems, light emitting diodes, and photovoltaic cells.

Flat panel display manufacturers—namely those that produce LCD panels used in products such as televisions, computer monitors, tablets, and mobile phones—use various F-GHGs and N_2O during panel production. These gases, which are highly effective in their process performance, are used to etch intricate patterns onto the glass, during deposition, and to rapidly clean the chemical vapor deposition (CVD) tool chambers [...] Fluorinated heat transfer fluids are also used in the manufacture of flat panel displays to cool manufacturing equipment [...]

During manufacturing of flat panel displays, a portion of F-GHGs pass through the manufacturing tool chambers unreacted and are released into the atmosphere [...] A portion of the F-GHGs used in processes may also react in chambers to form by-product emissions, or emissions of other F-GHGs. The magnitude of emissions can vary depending on a number of factors including: gas used, type/brand of equipment used, company-specific process parameters, and number of F-GHG using steps in a production process. Companies' manufacturing processes and, consequently, emissions also vary across flat panel manufacturing fabs (i.e., manufacturing, or fabrication, facilities).

Over the last decade, major flat panel suppliers have taken voluntary steps to reduce their F-GHG emissions. For example, in 2001, the World LCD Industry Cooperation Committee (WLICC) (including the LCD industry associations in Korea (the Environment Association of LCD in Korea or EALCD/EDIRAK), Taiwan (Taiwan TFT-LCD Association or TTLA), and Japan (the Industries Research Committee in Japan or LIREC/JEITA)), agreed to voluntary reduction activities and set a goal to reduce F-GHG emissions to 0.82 MMTCE by 2010. They estimated that these reductions represented one tenth of their anticipated emissions, effectively reducing 2000 baseline levels by approximately 90 percent (WLICC calculated that F-GHG emissions in 2010 would reach 8.2 MMTCE without implementing any reduction measures) [...] To meet the reduction goal, many suppliers in participating countries implemented strategies to address their emissions including installing abatement technologies on production lines in their newer generation fabs, namely those built within the last decade. As a result, F-GHG emissions were reduced by 10.1 MMTCE, to where aggregate emissions totaled 1.75 MMTCE. Though these reductions demonstrated significant accomplishments, the WLICC fell short of its goal due to a rise in emissions resulting from a rapid increase in production for LCD panels that were integrated into products such as televisions faster than initially anticipated [...]

Since the WLICC set its goals, newer suppliers with growing market share—those who have not participated in the WLICC's F-GHG reduction efforts to date—have also emerged and information on their F-GHG emissions reductions efforts is currently unknown. In addition, it appears that some key suppliers are still in varying stages of implementing comprehensive F-GHG emission reductions efforts across their fabs. As worldwide demand for flat panels, namely LCDs, continue to increase, F-GHG emissions are also projected to rise. To mitigate those emissions, it is important that reduction efforts across all major panel suppliers are implemented.

How Can F-GHGs Be Reduced?

Over the last decade, electronics manufacturers have made significant progress in identifying effective technological solutions to reducing F-GHG emissions. The following approaches to reducing F-GHG emissions resulting from flat panel manufacturing are in use today or are currently being explored:

1. Process improvements/source reduction: Manufacturers optimize their processes to use F-GHGs more efficiently, especially in the chemical vapor deposition clean processes, resulting in smaller amounts of gas that are unreacted and emitted.
2. Alternative chemicals: Manufacturers use alternative lower Global Warming Potential (GWP) or more efficient gases to accomplish the same result. In the case of chemical vapor deposition (CVD) remote plasma chamber cleaning, many manufacturers have modified their processes to be able to use NF_3 instead of SF_6. Though NF_3 still has very high GWP, it is lower than that of SF_6. Some companies are piloting the use of F_2 to replace NF_3 in the remote plasma chamber cleaning process, seeking to surmount some of the challenges associated with transport, storage, and use.
3. Capture and beneficial reuse: Manufacturers capture F-GHGs and process them to remove impurities and refine them for reuse. Some suppliers are evaluating the opportunities; reuse/recycling has so far not been implemented widely due to limitations on the effectiveness and cost of available technologies.
4. Abatement via gas destruction technologies: Both *point-of-use abatement,* where the abatement system is attached to tools, and *centralized abatement systems,* where gases are sent to, and destroyed in, a centralized system, are being used by major panel suppliers. The majority of abatement systems in use are combustion-based. Though suppliers employ a mix of strategies to reduce F-GHGs, abatement remains one of the most effective ways to reduce the majority of F-GHG emissions.

Measuring the efficiency of an installed abatement system to destroy or remove gases such as F-GHGs—known as the destruction or removal efficiency (DRE)—directly relates to how suppliers can account for their annual F-GHG emissions and subsequent reductions. Most suppliers today use default factors from the 2006 IPCC Guidelines to account for the DRE of abatement systems. However, suppliers may also directly measure DREs using measurement guidelines or protocols. An example of such a protocol is the United States Environmental Protection Agency's "Protocol for Measuring Destruction or Removal Efficiency (DRE) of Fluorinated Greenhouse Gas Abatement Equipment in Electronics Manufacturing" (EPA's DRE Protocol). Published in 2010 and internationally peer-reviewed, EPA's DRE Protocol provides a reliable method for measuring

DRE's of point-of-use abatement systems for F-GHGs used during the manufacture of electronics. In other cases, suppliers may monitor their systems on an ongoing basis, especially in the case of Clean Development Mechanism (CDM) projects, to acquire on-site real-time data. Suppliers may also test their abatement systems by monitoring specific parameters such as temperature, process gas, and exhaust gas flow rate. Going forward, EPA anticipates that additional information on industry practices can highlight which DRE measurement approaches are being used throughout the industry to produce reliable estimates of abatement systems' DREs.

Source: United States Environmental Protection Agency. 2016. "Center for Corporate Climate Leadership Sector Spotlight: Electronics." Overviews and Factsheets. September 22, 2016. https://www.epa.gov/climateleadership/center-corporate-climate-leadership-sector-spotlight-electronics. [website no longer active] Archived page available at https://19january2017snapshot.epa.gov/climateleadership/center-corporate-climate-leadership-sector-spotlight-electronics_.html.

"Countering WEEE Illegal Trade (CWIT) Summary Report, Market Assessment, Legal Analysis, Crime Analysis and Recommendations Roadmap" (2015)

Between 2013 and 2015 a consortium of United Nations agencies, Interpol, trade associations, and researchers conducted a research project focused specifically on the illegal trade of waste electrical and electronic equipment (WEEE) within and from Europe. The goal of the project was to quantify flows of WEEE within Europe and leaving its borders as well as to assess policy, regulatory, and procedural gaps in compliance and enforcement of existing rules and regulations. The results of the study offer some surprising insights that go against the grain of a good deal of the publicly expressed concerns around post-consumer e-waste. For example, despite the common concerns about exports of e-waste from rich countries in Europe to poor countries in Africa, the study found that ten times more e-waste was improperly handled within Europe itself than was exported from the continent to countries elsewhere. Another important finding from the project was that the reuse value of electrical and electronic equipment discarded in Europe as waste far exceeded the value of materials they contain when exported outside of the region.

That finding complicates the simplistic claim that electronics discarded as waste in richer markets are exported in order to avoid disposal costs. In other words, what the report found was that it is mainly reuse value driving exports rather than waste dumping.

The research undertaken by the Countering WEEE Illegal Trade (CWIT) project found that in Europe, only 35% (3.3 million tons) of all the e-waste discarded in 2012, ended up in the officially reported amounts of collection and recycling systems.

The other 65% (6.15 million tons) was either:

- exported (1.5 million tons),
- recycled under non-compliant conditions in Europe (3.15 million tons),
- scavenged for valuable parts (750,000 tons)
- or simply thrown in waste bins (750,000 tons).

1.3 million tons departed the EU in undocumented exports. These shipments are likely to be classified as illegal, where they do not adhere to the guidelines for differentiating used equipment from waste, such as the appropriate packaging of the items. Since the main economic driver behind these shipments is reuse and repair and not the dumping of e-waste; of this volume, an estimated 30% is e-waste. This finding matches extrapolated data from IMPEL on export ban violations, indicating 250,000 tons as a minimum and 700,000 tons as a maximum of illegal e-waste shipments.

Interestingly, some ten times that amount (4.65 million tons) is wrongfully mismanaged or illegally traded within Europe itself. The widespread scavenging of both products and components and the theft of valuable components such as circuit boards and precious metals from e-waste, means that there is a serious economic loss of materials and resources directed to compliant e-waste processors in Europe.

Better guidelines and formal definitions are required to help authorities distinguish used, non-waste electronic and electrical equipment (such as equipment coming out of use or in post-use storage destined for collection or disposal) from WEEE. Penalties must be harmonised to simplify enforcement in trans-border cases.

Organised crime is involved in illegal waste supply chains in some Member States. However, suspicions of the involvement of organised crime in WEEE are not corroborated by current information. Increased intelligence will lead to a more comprehensive understanding of the issue. Importantly, case analysis of illegal activities outlines that vulnerabilities exist throughout the entire WEEE supply chain (e.g., collection, consolidation, brokering, transport, and treatment). Offences include: inappropriate treatment, violations of WEEE trade regulations, theft, lack of required licenses/permits, smuggling, and false load declarations.

To address vulnerabilities more coherent multi-stakeholder cooperation is essential. For this purpose a recommendation roadmap with short, medium, and long term recommendations has been developed. These recommendations aim to reduce illegal trade through specific actions for individual stakeholders; to improve national and international cooperation to combat illegal WEEE trade, actions such as:

- Increasing involvement, and improving awareness of users in the early stages of the e-waste chain;
- An EU-wide ban on cash transactions in the scrap metal trade;
- Mandatory treatment of WEEE according to approved standards, and dedicated mandatory reporting of treatment and de-pollution results;
- Better targeting, more upstream inspection, and national monitoring;
- An Operational Intelligence Management System (OIMS) to support intelligence-led enforcement and identify the risks associated with organised crime groups;
- A National Environmental Security Task Force (NEST), formed by different authorities and partners, to enable a law enforcement response that is collaborative and coordinated at national, regional, and international level; and
- Dedicated training of judges and prosecutors [...]

In total, 1.5 million tons [of WEEE] are leaving the EU. 200,000 tons are documented as UEEE [used electrical and electronic equipment] exports. This figure is based on more detailed mass balances for five high income countries and covers the highest value portion of the export for reuse totals; being relatively well-tested and functioning (often IT) equipment. These devices typically have considerable remaining lifetime and thus reuse value and are commonly covered for example by professional refurbishers and/or charity organisations donating well-tested computers to educational institutes in Africa. This flow is most likely also occurring for other rich EU countries, however this could not be quantified in this project.

The remaining 1.3 million tons is also predominantly UEEE, but is frequently mixed with WEEE and repairable items. The entire amount is a grey area subject to different legal interpretations and susceptible to export ban violations. At some point in these reuse activities; the originally discarded WEEE is no longer regarded as waste. This occurs where the items are refurbished, tested and properly packed for export.

However, the entire amount is a grey area since there are many more issues besides the distinction between WEEE versus UEEE. Shipments often include parts, functioning but very old UEEE with no real value or market anymore, or with very short remaining lifespans as well as WEEE which is repairable, and relatively new but non-functioning appliances ideal for harvesting of spare parts, etc. In any case, many shipments are not following the existing guidelines as sorting, testing and packaging in Europe comes at a cost.

The quality of a large part of these shipments of products needs to improve. The remaining 1.3 million tons (based on the most recent literature sources, and combined with inspection observations) is estimated to consist of around 70% as functioning second-hand items (900,000 tons) and 30% of WEEE (400,000 tons), including repairable items […]

In short, mismanagement of discarded electronics within Europe involves ten times the volume of e-waste shipped to foreign shores in undocumented exports […]

To what extent does the mismanagement of volumes that occurs all along the WEEE chain damage the environment and the European economy at large? How does this affect the EU's vision to turn the linear economy into a circular economy?

In this respect, it should be noted again that the main driver behind exports is the reuse value combined with the avoided costs of sorting, testing and packaging. The economic values of the exports cannot be quantified in detail because there is no clear information. The exports involve too many individual appliance types and different price levels in the receiving countries. The Environment Agency in the UK provides an example of a typical profit value of £8,000 for a container of mixed, unsorted and untested equipment sent to Africa. This indicates that the magnitude of the reuse value is multiple times the material value of the contents.

Source: Huisman, Jaco, I. Botezatu, L. Herreras, M. Kiddane, J. Hintsa, V. Luda di Cortemiglia, P. Leroy, et al. 2015. "Countering WEEE Illegal Trade (CWIT)

190   *Electronic Waste*

Summary Report, Market Assessment, Legal Analysis, Crime Analysis and Recommendations Roadmap." Lyon, France.

"Future E-waste Scenarios," United Nations University/ United Nations Environment Programme (2019)

This report was published by a consortium of researchers with support from the Solve the E-waste Problem (StEP), the United Nations University (UNU), and United Nations Environmental Program International Environmental Technology Centre (UNEP IETC). The research attempts to envision different alternative scenarios for e-waste up to the year 2030. The report is useful for its description of three future scenarios each with their attendant characteristics related to technology, impacts, management, policies, and business model. That the report identifies multiple possible future scenarios for e-waste is helpful for understanding that there is no necessary linear trajectory from today to any one future about pollution and waste arising from electronics. Even the three scenarios envisioned in this report are not exhaustive. Readers may find some inspiration in the multiplicity of futures described in the report to describe additional future alternatives beyond what this report imagines.

StEP [Solve the E-waste Problem], UNU [United Nations University], and UNEP IETC [United Nations Environmental Program International Environmental Technology Centre] have been working extensively on e-waste issues and made an attempt to look into the future of the problem in order to initiate policy level discussions on the challenges and opportunities ahead. Having insight into the future will help policymakers and industries, as well as other stakeholders, to make better strategic decisions. Forecasting is also necessary vis-à-vis strategic concepts towards sustainable development, such as circular economy and the UN's Agenda 2030.

We cannot expect immediate success with these concepts without an active search solutions. The complicated nature of production, use, and disposal of electronics require significant changes in order for the processes to become sustainable.

[...]

As technological advances continue, the use of e-products in people's daily lives will only grow. Although future developments are impossible to predict precisely, speculations can be made based on past experiences and current trends in this sector. [...] three scenarios can be imagined: 1. Linear Growth, 2. Reactive Approach, 3. Proactive Pathway.

SCENARIO 1: *LINEAR GROWTH*

The business-as-usual scenario, where a standard growth-based economic agenda is the priority, continues, resulting in a Linear Growth scenario. The consumption of e-products and the amount of e-waste grow at the usual rates. Conventional business models remain dominant with only a few exceptions. E-waste management capacities are also lagging behind.

All these factors combined result in an increased consumption and a severe e-waste problem.

TECHNOLOGY

- The pace of e-product technological innovation continues but little is achieved in terms of breakthroughs for making e-products' lifecycle more sustainable
- Ongoing market segmentation results in less wealthy markets, allowing lower-tier manufacturing of low-quality products (with consequences for resource use and toxicity)
- Products are designed and built for demand in growing economies, which results in cheaper but low-quality products with shorter life
- Consumer e-products increasingly become disposable, "single-use" quality, but e-waste recycling is limited due to a combination of material heterogeneity and cost of recycling

IMPACTS

- The growing use of critical resources and fossil-based polymers increases environmental footprint of e-products
- Toxicants are still used in manufacturing processes, which are occasionally released to local environments, and the toxic substances also pose challenges at the product EoL
- The development inequality between rich and poor countries makes it even more difficult to control e-waste exports to places without proper management systems

MANAGEMENT

- EoL [End of Life] systems are designed mainly to minimize costs but they fail to improve e-waste collection and subsequent recycling
- The potential for reuse of products and components are not utilized
- Material recycling is driven by economic interests, in both informal and formalized settings
- "Cherry-picking" of valuables and disposal of unrecyclable items and hazardous substances continues despite formalization of e-waste recycling in developing countries
- Lack of collaboration among stakeholders hinders possible optimization of resource use

POLICIES

- No significant changes in e-waste policies, product design, or circular economic policy
- Although the existing regulations have set targets for collection and reporting for EoL products, they are not strictly implemented
- Product design does not improve because Eco-Design initiatives are not implemented effectively
- Robust implementation of export bans is lacking, resulting in illegal flows of e-waste
- Toxicants in new products decline as manufacturers conform to RoHS-like legislations, but aggregate toxicity never fully dissipates as global demand for EEE rises

BUSINESS

- Market is flooded with several cheaper options for e-products that do not reflect the true/environmental cost of production
- The core idea of EPR [extended producer responsibility] system is limited to sales and collection data reporting; no physical "take-back" takes place, hence no feedback loop for improving product design
- Reuse, lifetime extension, and alternative business models (e.g., leasing) are tried only by a few companies for select product types (that are profitable)
- Conventional consumption models that continue to dominate are sales-driven with no consideration for lifecycle impacts of e-products or recyclability

USERS

- Users are unaware of impacts linked to production, use, and EoL management of e-products
- Not many users know how they can help improve the situation, for example by purchasing better-designed products and facilitating e-waste collection and recycling
- Reuse and repair are not mainstream options, and many users are buying new items
- Even informed people fail to act due to technological, social, and economic barriers

SCENARIO 2: *REACTIVE APPROACH*

Strong regulations and monitoring frameworks are in place. Businesses are reluctantly taking the Reactive Approach to comply with the new set of legislations. As a result, some changes appear in the production and consumption patterns. This helps to tackle the localized issues linked to manufacturing and ewaste management, mainly in developed countries.

But the changes in the global practices of production and EoL management of e-products are slower.

TECHNOLOGY

- The technological development and use of e-products continue to grow with little change
- "End-of-pipe" recycling technologies develop, driven mainly by the value of secondary resources
- Advanced sorting and material recycling techniques make it possible to recirculate resources, creating closed-loop supply chains
- As the demand and price of some critical resources increase, material substitution is a priority for producers

IMPACTS

- Strict policies improve the situation in leading economies but the issues linked to EoL management (informal handling, material losses, and pollution) in emerging markets persist

194   *Electronic Waste*

- Regulatory push for substantial reduction and elimination of toxicants, in manufacturing and in final products, helps to achieve per unit improvements. However, the aggregated risk of toxicity persists with the growing demands of e-product

MANAGEMENT

- Higher e-waste collection and reuse targets for all product groups are set and are being slowly implemented
- Material recovery targets for critical resources from e-waste are also introduced
- The EU has begun to issue fines to member states who fail to achieve these targets, which puts pressure on businesses operating in the region
- The implementation of EPR-based e-waste management is not yet effective in countries with newly introduced EPR legislations

POLICIES

- A patchwork of EPR regulations have arisen across the world. Policy response continues along the same line but in a more extreme direction for collection & resource recovery targets
- E-products are required to have longer warranty period, which has forced producers to design products for durability and easy repair
- Green Public Procurement is now mandatory for all public purchases in Europe and further incremental changes to the criteria are being incorporated
- WEEE, EcoDesign, RoHS, and associate policies are de facto regulatory frameworks globally

BUSINESS

- Producers adapt, more or less grudgingly, to policy frameworks shifting. This helps to improve the consumption pattern and encourages businesses to adopt more circular models.
- Specific legal requirements (e.g., spare parts and software updates, ease of repair, replaceable batteries, and standardized components) are incorporated into e-products
- Businesses are obliged to be responsible for their EoL products in countries with no EPR legislations

USERS

- The cost of compliance to these measures forces companies pass those costs to users, and thus depresses sales in established markets and delays launches in emerging markets
- Users are still not ready to change their consumption habits mainly because of lacking awareness and little financial motivation
- Hoarding used products in people's homes and incorrect disposal remains frustratingly high and makes EoL targets more expensive to achieve

SCENARIO 3: *PROACTIVE APPROACH*

This utopian world envisions a drastic reduction in consumption in which businesses are choosing the Proactive Path toward sustainable production and consumption practices. Businesses develop more sustainable consumption practices along the e-products' supply chain, which are supported by governments and accepted by users. All stakeholders, including economic actors, are supporting the commitment from producers to take a lifecycle approach in manufacturing and EoL management of e-products.

TECHNOLOGY

- With a shift toward renewable energy and use of batteries, there is a growing demand of some resources (e.g., Lithium and Cobalt)
- Products are designed to last longer, to facilitate easy maintenance during use, and for the ease of dismantling and recycling at the EoL
- Technologies like 3D printing help product-lifetime extension by making it possible for users to create spare parts
- The EoL resource-recovery technologies better meet the need of e-waste

IMPACTS

- Although the amount of e-waste generated increases, the management capacity ensures environmentally-sound practices with little impacts
- Reuse of e-products and their components, along with recycling of e-waste, helps to control the demand of virgin resources
- Responsible mining practices are more common with less social and environmental issues

Significant reductions in toxicants entering the environment from e-products manufacturing

MANAGEMENT

- EoL management is an integral part of product development and concepts such as "design for recycling" are part of design checklist
- Producers and EoL managers (e.g., collectors, repair shops, recyclers) are working together to optimize EoL resource recovery from consumer products
- Following circular economy principles, priorities are put in place to facilitate the reuse of products and components before material recycling

POLICIES

- Policies are based on ambitious goals of achieving truly circular systems and businesses are leading the process of drafting and implementing progressive legislation
- Regulations function as a guidance rather than a compliance instrument
- Less resource-intensive products and consumption models are encouraged through rewards and recognition
- All countries are covered by legislation that addresses all lifecycle stages of e-products, including e-waste management

BUSINESS

- Businesses are the leaders of a shift toward sustainable production and consumption practices with due diligence
- Products, services, and supply chains are designed to facilitate a closed loop of materials
- Businesses also enjoy the potential savings from resource reuse
- New technology and innovation make this possible, along with the growing acceptance of business models that discourage consumers to privately own several e-products
- Thanks to information campaigns led by grassroots initiatives, businesses, and governments, users are aware of the e-waste problem and the environmental footprints of e-products

USERS

- User behavior is considered when designing e-waste collection systems to create more functional infrastructure and achieve higher collection and recycling rates
- Alternative business models (e.g., leasing and product-service systems) are becoming "mainstream," which is made possible by the proactive industries
- More people are opting to repair broken products—thanks to repair businesses and grassroots initiatives

OPPORTUNITIES

Although there are uncertainties about how future technology will evolve, the use of e-products—and thus the generation of e-waste—will almost certainly grow, at least during the next few decades. It is especially true for rapidly growing economies that are yet to be flooded with the myriad of e-products that come with economic prosperity. It means more challenges, as well as opportunities, for all players in the global e-waste arena: producers, users, e-waste collectors, recyclers, and policy makers. Regardless of which direction the future evolves, ensuring a sustainable production and consumption system for e-products will require significant efforts from all stakeholders.

With insights into the future, producers probably have the best opportunity to design a future-proof electronics sector that is sustainable economically as well as environmentally. Besides manufacturing and recycling, innovative businesses can also tap into the huge product and component reuse potential. This can create local businesses and help countries, especially those with no reserves for primary resources or e-products manufacturing, to utilize the functional value of products for a longer period, avoiding the import of new items. Many products ranging from household appliances (washing machine and lighting equipment) to ICT equipment (mobile phones and computers) can be offered as a service or leased instead of selling products. Such business models can provide better opportunities for product lifetime extension and smoother take-back at the EoL.

Exercising due diligence, businesses can minimize the potential challenges of critical resources and stricter legal requirements. Some big brands are hitting the limits of business models that rely on sales of new products, whereas others who prioritize product longevity are gaining popularity. There is also demand for more sustainable products, which can be attributed to growing consumer awareness. Businesses will be better off by proactively addressing users' demand.

Source: Parajuly, Keshav, Ruediger Kuehr, Abhishek Kumar Awasthi, Colin Fitzpatrick, Josh Lepawsky, Elisabeth Smith, Rolf Widmer, and Xianlai Zeng. 2019. "Future E-Waste Scenarios." Bonn: United Nations University/ United Nations Environment Programme. https://collections.unu.edu/eserv/ UNU:7440/FUTURE_E-WASTE_SCENARIOS_UNU_190829_low_screen.pdf.

"Nixing the Fix: An FTC Report to Congress on Repair Restrictions," United States Federal Trade Commission (2021)

In 2019 the United States Federal Trade Commission (FTC) hosted a workshop called, "Nixing the Fix: A Workshop on Repair Restrictions." An important impetus of the workshop was growing organized citizen and consumer action to resist what advocates argued was unfair, if not outright illegal, behavior on the part of original equipment manufacturers (OEMs) to thwart independent and do it yourself repair, especially of electronic devices. A major issue of the FTC's investigations and the workshop involves the degree to which OEMs are violating the Magnuson-Moss Warranty Act (MMWA). There are many important nuances to interpreting the MMWA, but one particular provision within the legislation is especially important: Section 102(c), also known as the anti-tying provision. Essentially this provision blocks OEMs from invoking warranty coverage as a way to negate consumers' right to repair their own devices or to have them repaired by an independent third-party, including the use of non-OEM replacement parts. Although OEMs put forward a number of reasons why they restrict independent repair, the FTC found that most of those reasons were not supported by the evidence. Repair of electronics is an important intervention into mitigating the pollution and waste arising from the mining for and manufacturing of them. Repair is a practice that acts as a form of energy and material conservation.

EXECUTIVE SUMMARY

[...]

Congressional interest in the competition and consumer protection aspects of repair restrictions is timely. Many consumer products have become harder to fix and maintain. Repairs today often require specialized tools, difficult-to-obtain parts, and access to proprietary diagnostic software. Consumers whose products break then have limited choices.

Furthermore, the burden of repair restrictions may fall more heavily on communities of color and lower-income communities. Many Black-owned small businesses are in the repair and maintenance industries, and difficulties facing small businesses can disproportionately affect small businesses owned by people of color. This fact has not been lost on supporters of prior right to repair legislation, who have highlighted the impact repair restrictions have on repair shops that are independent and owned by entrepreneurs from underserved communities. Repair restrictions for some products—such as smartphones—also may place a greater financial burden on communities of color and lower-income Americans. According to Pew Research, Black and Hispanic Americans are about twice as likely as white Americans to have smartphones, but no broadband access at home. Similarly, lower-income Americans are more likely to be smartphone-dependent. This smartphone dependency makes repair restrictions on smartphones more likely to affect these communities adversely.

The pandemic has exacerbated the effects of repair restrictions on consumers. [...] the pandemic has made it harder to get broken devices fixed, as many big chain stores have ceased offering on-site repairs. As a result, people have been forced to send their devices to authorized repair facilities—often waiting weeks for them to be returned.

The pandemic also has revealed a drastic shortage in the availability of new laptops for students. An Associated Press examination of the availability of school laptops found that the increased demand for computers and supply chain challenges posed by the pandemic had resulted in laptop shortages in school districts around the country. For instance, California has reported the need for 1 million laptops for students and Alabama was waiting on 33,000 student computers. Kinks in the semiconductor supply chain are now posing an additional threat to the supply of new laptops. Reducing barriers to repair may permit older laptops to be refurbished more easily, thereby expanding the supply of available laptops [...]

[...] many manufacturers restrict independent repair and repair by consumers through:

- Product designs that complicate or prevent repair;
- Unavailability of parts and repair information;

- Designs that make independent repairs less safe;
- Policies or statements that steer consumers to manufacturer repair networks;
- Application of patent rights and enforcement of trademarks;
- Disparagement of non-OEM parts and independent repair;
- Software locks and firmware updates; or
- End User License Agreements.

Manufacturers explain that these repair restrictions often arise from their desire to protect intellectual property rights and prevent injuries and other negative consequences resulting from improper repairs [...]

CONCLUSION

The debate around repair restrictions illustrates the limitations of MMWA's [Magnuson-Moss Warranty Act] anti-tying provision in repair markets. While the anti-tying provision gives consumers the right to make repairs on their own or through an independent repair shop without voiding a product's warranty, repair restrictions have made it difficult for consumers to exercise this right. Although manufacturers have offered numerous explanations for their repair restrictions, the majority are not supported by the record.

Source: United States Federal Trade Commission. 2021. "Nixing the Fix: An FTC Report to Congress on Repair Restrictions." https://www.ftc.gov/system/files/documents/reports/nixing-fix-ftc-report-congress-repair-restrictions/nixing_the_fix_report_final_5521_630pm-508_002.pdf.

6

Resources

This chapter provides an annotated list of selected books, journal articles, and reports that cover pollution and waste arising from electronics from a variety of perspectives. Several key peer-reviewed publications are also highlighted. These peer-reviewed publications include numerous relevant research articles on waste and pollution issues associated with the mining for, the manufacturing of, the use of, and the management of end-of-life electronics. Much of the peer-reviewed research published on these topics typically falls in the domains of various environmental sciences and engineering disciplines, which is what is highlighted in this chapter. However, searches in the peer-reviewed humanities and social sciences literature can also surface relevant articles.

Books

Gabrys, Jennifer. 2011. *Digital Rubbish: A Natural History of Electronics*. Ann Arbor: University of Michigan Press.

Digital Rubbish offers a scholarly assessment of digital discards from a humanities perspective. In this respect, the book provides a more expansive understanding of e-waste than just a focus on post-consumer discards. The text draws on the work of cultural critic Walter Benjamin's idea of natural history which understands discarded objects and other detritus as a paleontologist might approach fossils—that is, as physical remnants of whole ways of being most or all of which have otherwise disappeared and leave behind only fragments

---

An electronics recycling event in Georgia, USA. Post-consumer recycling garners significant attention as a solution to e-waste reduction. Although its effectiveness in tackling the problem of e-waste as a whole is limited, recycling is still worthwhile in the effort to attack the problem from all angles. (Bluiz60/Dreamstime.com)

from which to glean broader understanding and meaning. In this book, digital rubbish takes forms that are both expected and unexpected. It unearths the pollution and waste arising from electronics manufacturing in Silicon Valley but also a broader meaning of digital discards that museums and archives must deal with as both digital media and the machines they run on become obsolete at rates much faster than sheaves of paper, microfilm, or books. In *Digital Rubbish*, the remainders of digital technology—from pollutants soaking the ground of Silicon Valley to "junk bonds" traded on electronic stock exchanges—are shown to be important physical remnants, but also allegories that analysts can unpack for their moral and political lessons.

Grossman, Elizabeth. 2006. *High Tech Trash: Hidden Toxics and Human Health.* Washington, DC: Island Press.

Grossman's book is a work of environmental journalism. It is an accessible critical examination of the lifecycle of electronics from mining to manufacturing to disposal. *High Tech Trash* is written for, and marketed to, a primarily US audience and the facts and figures used throughout the text reflect that emphasis. The book addresses the international trade and traffic of electronic waste primarily through the mode of qualitative storytelling. The penultimate chapter of Grossman's book is an eye-opening descriptive account of the gritty realities of electronics recycling in industrial systems. In the book's final chapter Grossman builds on Aldo Leopold's idea of a land ethic to offer more holistic solutions to pollution and waste arising from electronics.

Hieronymi, Klaus, Ramzy Kahhat, and Eric Williams, eds. 2013. *E-Waste Management: From Waste to Resource.* New York: Routledge.

The editors of this collection include the chairperson of Hewlett Packard's environmental board and two sustainability engineers. The chapters are written by authors in a variety of engineering disciplines spanning a range of sub-disciplines including electronic engineering, materials science, and process engineering. The book is oriented toward a waste management perspective with an emphasis on technical solutions to e-waste problems. Chapters cover such topics as current and future electronic waste recycling technologies, the management of critical metals, the management of plastics through circular economy systems, assessments of recycling systems in North America and Europe, current and future international trade in electronic waste, the role of reuse in mitigating post-consumer electronic discards, and future trends in managing electronic scrap. The collection contains useful quantitative analyses of different material flows derived from e-waste and transboundary flows of second-hand electrical and electronic equipment as well as electronic scrap.

Kuehr, Ruediger, and Eric Williams, eds. 2003. *Computers and the Environment: Understanding and Managing Their Impacts.* Eco-Efficiency in Industry and Science; v. 14. Dordrecht: Kluwer Academic Publishers.

This edited collection is one of the earliest book-length treatments of the environmental impacts of computing equipment as a class of manufactured products. At the time of publication both editors were based at the United Nations University (UNU). UNU would go onto lead or support much of the primary research and policy analysis undertaken by UN bodies seeking to mitigate or eliminate waste and pollution from electronics, particularly from post-consumer discarded devices.

Computers in the Environment is a notable collection on several fronts. It is, for example, one of the few sources of substantive analysis on pollution and waste arising in the manufacturing phase of electronics. The latter phase coupled with mining is the most impactful stage of the lifecycle of electronics from a pollution and waste perspective. Much of the analysis in this book directed at the manufacturing stage in the lifecycle of electronics documents and provides data on the energy and material requirements for manufacturing personal computers and other consumer electronics. The book also addresses topics that would later become more familiar in assessments of the e-waste problem such as the implications of regulations being promulgated in European Union around recycling waste electrical and electronic equipment. The book is also notable for chapters on the market for used personal computers and policy prescriptions for supporting that market in ways beneficial to reuse. The latter represents a prescient analysis of a theme that would later become prominent in the research literature around the e-waste problem especially including the roles that reuse and repair can play in material and energy conservation. Despite being an older publication, this book represents in many ways productive analytical roads not taken by activists and researchers concerned with e-waste understood too narrowly as an end-of-pipe, post-consumer problem.

206 *Electronic Waste*

Lepawsky, Josh. 2018. *Reassembling Rubbish: Worlding Electronic Waste.* Cambridge, MA: MIT Press.

Reassembling Rubbish is premised on a seemingly simple question: what is the right thing to do with electronic waste? The book then unpacks the complex conceptual and empirical implications of that question. Lepawsky draws on insights from the fields of geography and science and technology studies to demonstrate that e-waste is simultaneously a problem of knowledge and of what he calls 'worlding"—a concept used to describe the practical action of people, places, and things working in the world and toward the world they think it ought to be. In practical terms, Lepawsky demonstrates through a combination of quantitative and qualitative evidence that the typical solutions put forward to solve the problem of e-waste, such as recycling, cannot scale to match the magnitude of overall pollution of waste arising from electronics. The final chapter of the book examines the prospects of several conventional and less conventional possible solutions for problems of pollution and waste arising from the mining, the manufacturing, use, and discard of electronic devices.

Little, Peter C. 2014. *Toxic Town: IBM, Pollution, and Industrial Risks.* New York: NYU Press.

Toxic Town is an ethnographic examination of the legacy of toxic pollution at one of the key "birthplaces" of the high-tech age—Endicott, New York—where IBM built its first production plant in 1924. *Toxic Town* provides a rich and critical analysis from an anthropological perspective on one of the key sites where legacy pollution arising from electronics manufacturing continues to have harmful consequences for residents and former employees of IBM in a specific place. Using ethnographic methods, Little weaves an analysis of how real people learn to live with toxic risks, navigate scientific knowledge and uncertainty, and advocate for themselves and their community in the face of corporate power.

Little, Peter C. 2022. *Burning Matters: Life, Labor, and E-Waste Pyropolitics in Ghana.* New York: Oxford University Press.

In *Burning Matters* Little brings an anthropological ethnographic analysis to e-waste management in Ghana. Much of the book is focused on situating life and work at Agbogbloshie—a site frequently characterized as the largest e-waste dump in the world before it was cleared in 2021—within broader national and international cultural, economic, and environmental networks with a particular focus on how people who work with e-waste at the site manage the risks of negative health outcomes with their need to make a livelihood.

Little, Peter C. 2023. *Critical Zones of Technopower and Global Political Ecology: Platforms, Pathologies, and Plunder*. Lanham, MD: Lexington Books.

Little's *Critical Zones* is a conceptual and empirical synthesis of much of his previous work on IBM in Endicott, New York, and e-waste processing in Ghana. This synthesis is put to work and extended toward a broader political ecology critique of "big tech." Political ecology is a wide field of investigation that cuts across many disciplines. At a basic level it approaches nature, politics, economics, and society not as separate interacting variables but as inherently enmeshed with one another such that an understanding of any one of those categories is dependent on understanding the others as well. Little argues that a crucial task is developing a situated, but global analysis of the environmental health consequences of pollution and waste arising from electronics in the wake of the ongoing rise of big tech. The book is divided into three sections organized by theme: power, health, and environment.

Maxwell, Richard, and Toby Miller. 2012. *Greening the Media*. Oxford, UK; Toronto: Oxford University Press.

Greening the Media is an accessibly written, critical analysis directed primarily toward media studies scholars with activists, students, and a more general concerned readership also in mind. The central purpose of the book is to undo the prevailing notion, prevalent across a broad range of discourses, that "new media" supported by electronic networks are immaterial conduits for an environmentally clean and light "information age." The authors insist that digital media demand a materialist analysis that more fully accounts for its environmental implications at every stage of digital media's lifecycle.

Maxwell, Richard, Jon Raundalen, and Nina Lager Vestberg. 2014. *Media and the Ecological Crisis*. New York: Routledge.

This edited collection compiles essays that deepen and enrich the project of Maxwell and Miller's book, *Greening the Media*. *Media and the Ecological Crisis* provides an account of media technologies that eschew the presumption that because media are digital, they are not also material. Like Maxwell and Miller's book, *Media and the Ecological Crisis* is written for media studies scholars, practitioners, and activists. A chapter by Jennifer Gabrys uses the case of "smart" energy meters to develop the concept of "electronic environmentalism" which helps to grasp the paradox of the increasingly common attempts to regulate consumptive behavior (e.g., energy use) using digital technologies that are themselves energy intensive to manufacture and run. Gabrys's chapter brings important attention to the environmental implications of the manufacturing and use of electronics, which extends the analytical frame around e-waste beyond treating it as only that which arises when consumers discard devices. In another chapter, Sophia Kaitatzi-Whitlock expands the narrower meaning of e-waste typically encountered in dominant storylines about it to include the ballooning "useless" information that accompanies the expansion of the "information economy" and the concomitant surplusing of skilled humans as they are replaced by algorithmic content generation. *Media and the Ecological Crisis* is a useful source for better understanding why a wider frame of analysis around e-waste is critical to more fully understanding the environmental consequences of these technologies.

O'Neill, Kate. 2019. *Waste.* Cambridge, UK; Medford, MA: Polity Press.

O'Neill's book is about the international political economy of waste and waste trading. International political economy, broadly speaking, understands economic relationships to be shaped by politics and, vice versa, politics to be shaped by economic exchange. As such, an important purpose of O'Neill's book is demonstrating how the complex relationships between waste and resource are organized and struggled over within a globalized economy. To do this work, the book devotes chapters to the rise of the global waste economy, conceptualizing waste as resource, and an examination of formal and informal livelihoods that work with wastes and convert them into resources. Much of the book is devoted to three case studies of particular waste categories to examine empirically the complexity of the international political economy of waste as resource. Discarded electronics is one of these case studies. The chapter devoted to them demonstrates the complexity of grappling with measuring e-waste, its tricky existence as both waste and resource, how and why the waste is traded across borders, and discussion of repair and reuse as examples of valuable labor done by people working in informal economies. Although only one chapter of O'Neill's book is devoted to discarded electronics, it is a valuable analysis due to the comparisons and contrasts that the book offers with other case studies of waste-to-resource in the global economy, namely, food waste and plastic scrap.

Siegel, Lenny, and John Markoff. 1985. *The High Cost of High Tech: The Dark Side of the Chip.* Toronto: HarperCollins.

Siegel and Markoff's volume is an early critical treatment of the social and environmental costs of the electronics industry. The book frames its analysis from the point of view of the role of electronics in the international geopolitics of the arms race and President Reagan's weapons programs of the 1980s. Chapter 8, called "Toxic Time Bomb," deals with pollution and hazardous waste risks associated with electronics manufacturing. Much of the chapter is devoted to the plume of toxic chemicals found to be leaking from chip manufacturing plants in the Silicon Valley region. The chapter is a good example of an early literature on pollution and waste from electronics that later framings of e-waste as an end-of-pipe, post-consumer problem overlooked.

Smith, Ted, David Allan Sonnenfeld, and David N. Pellow. 2006. *Challenging the Chip: Labor Rights and Environmental Justice in the Global Electronics Industry*. Philadelphia: Temple University Press.

This edited book remains one of the most definitive analyses of labor and environmental issues throughout the lifecycle of electronics. The book comprises eighteen chapters written by academics, activists, and journalists and is divided into three main parts. Part I offers an introduction to the changing global distribution of electronics manufacturing, discussions of occupational health and safety in electronics manufacturing, the gendered dynamics of labor migration associated with the industry, and several case studies of the sector in China, Thailand, India, and central and eastern Europe. Part II provides an overview of environmental justice and labor rights in the sector. These issues are examined through detailed case studies of Silicon Valley, Greenock, Scotland (the "Silicon Glenn"), the US-Mexico border and the *maquiladora* system, and Taiwan's electronics manufacturing zones. Part III turns to the issue of electronic waste and extended producer responsibility. This part of the book covers pollution and waste arising from the sector from the point of view of design and manufacturing as well as post-consumer discarding.

Articles and Reports

Mining

Deberdt, Raphael, and Philippe Le Billon. 2021. "Conflict Minerals and Battery Materials Supply Chains: A Mapping Review of Responsible Sourcing Initiatives." *The Extractive Industries and Society*, May 2021. https://doi.org/10.1016/j.exis.2021.100935.

"Conflict minerals" is an umbrella term sometimes applied to a suite of metals including tin, tungsten, tantalum, gold, cobalt, graphite, lithium, manganese, and nickel particularly when sourced from regions in Africa experiencing violent conflict between various groups of military and irregular armed forces. The underlying reasons for these conflicts are complex and variable, but there is evidence that control of mineral sources and mining are an important source of financing for these different groups. Conflict minerals are important for manufacturing a broad suite of electronic devices or products that include electronic components, including everything from consumer electronics such as phones and laptops to batteries for electric vehicles to renewable energy infrastructure such as windmills and solar panels. Concerns about the conditions associated with conflict minerals focus on labor conditions, especially the issue of child labor, political corruption, and environmental degradation. Mining is by far the single greatest source of pollution and waste associated with electronics.

In this article, Deberdt and Le Billon offer a systematic review of 220 studies of responsible mineral sourcing. Their goal is to identify the differing approaches that various responsible mineral sourcing initiatives take. Temporally the authors focus on studies published between 2010 and 2020 following the passage of the Dodd-Frank act in the United States, a provision of which requires companies to disclose whether they sourced conflict minerals from regions in Africa. Among the authors' findings are that particular minerals receive far more attention in these sourcing initiatives than a full range of minerals that can fall under the conflict mineral umbrella. Most of the attention is devoted to so-called 3TG minerals (tantalum, tin, tungsten, and gold) followed by cobalt. However, which minerals receive the most focus changes over the decade of studies analyzed by Deberdt and Le Billon. While 3TG saw growing attention between 2010 and 2019, cobalt becomes a matter of concern beginning in 2015 and continues to grow to 2020. Meanwhile, attention to lithium receives a bump of attention beginning in 2016 but appears to take off after 2019. What these authors' research suggests is that particular minerals become symbolic of much more general concerns for responsible mineral sourcing by different interests such as brand manufacturers, environmental NGOs, and civil society groups, particularly those located external to the site of the conflicts themselves. There is also a suggestive symbolic link between particular minerals and specific technologies in which they are used. For example, the attention to lithium grows as broader conversations about electric vehicles become prominent.

Fitzpatrick, Colin, Elsa Olivetti, T. Reed Miller, Richard Roth, and Randolph Kirchain. 2015. "Conflict Minerals in the Compute Sector: Estimating Extent of Tin, Tantalum, Tungsten, and Gold Use in ICT Products." *Environmental Science & Technology* 49, no. 2: 974–81. https://doi.org/10.1021/es501193k.

This article measures actual physical quantities of tantalum, tin, tungsten, and gold (3TG) found in a sample of five categories of information and communication (ICT) devices: servers, desktops, displays, laptops, tablets, and smartphones. These quantities are then scaled up to global estimates using publicly available device shipment data to estimate the total value of 3TG minerals attributable to what the authors call the compute sector. The results of the research offer some surprising findings given the attention that electronics and conflict minerals receive. At the time of study (2013), the authors' model suggests that the upper range of each mineral that can be globally attributed to the compute sector is 0.1 percent for tungsten, 2 percent for tin, 3percent for gold, and 15 percent for tantalum. The authors acknowledge some important limits to their study including that the five categories of ICT devices they study are not inclusive of all electronics. Therefore, they acknowledge that their analysis underestimates the total percentages of 3TG minerals that can be attributed to the compute sector.

Notwithstanding these limitations, however, the authors' results demonstrate the importance of thinking carefully about scalar mismatch. When activist and policy concerns about conflict minerals and responsible mineral supply chains focus too narrowly on a charismatic sector, such as electronics, which, though it is populated by many highly recognizable brands, advocacy for changing the organization of supply chains based on that sector may have unintended consequences. For example, if such advocacy actually resulted in the entire computer sector sourcing tantalum minerals from regions deemed to not be experiencing conflict, then at most 15 percent of global consumption of tantalum would be accounted for while leaving 85percent of such sourcing unchanged. The precision of these numbers can certainly be debated, but the broader issue of aligning proposed solutions with the magnitude of the problem so framed needs to be carefully considered.

Global Witness. 2022. "The ITSCI Laundromat: How a Due Diligence Scheme Appears to Launder Conflict Minerals." https://www.globalwitness.org/documents/20347/The_ITSCI_Laundromat_-_April_2022.pdf.

212 *Electronic Waste*

Global Witness is an international NGO focused on mitigating and eliminating exploitative supply chains for resources. In 2022, the organization published an investigative report assessing the effects of a trade association's attempt to set up a responsible mineral sourcing initiative for tin extracted from the African Great Lakes region, especially the Democratic Republic of Congo (DRC) and Rwanda. That initiative is called the International Tin Supply Chain Initiative (ITSCI). The goal of ITSCI is to provide assurances to downstream purchasers of tin mined in the African Great Lakes region that given shipments of the mineral are not associated with child labor, smuggling, or violent conflict in the region.

Global Witness conducted field research in the areas where ITSCI's assurance system is in operation. These researchers found that significant volumes of minerals certified by the ITSCI system contain substantial percentages of those minerals sourced from mines outside the ITSCI system. Consequently, the Global Witness report argues that ITSCI's system does not provide assurance that purchases of tin from the region are, in fact, free of undesirable associations. The Global Witness report makes thirty recommendations to mitigate or eliminate the problems found with the ITSCI system. These recommendations are directed toward various government actors in the region and beyond as well as downstream companies that ultimately use tin mined from the African Great Lakes region. The number of recommendations, and the broad range of actors to whom they are directed, are indicative of the complexity of the conflict mineral issue.

United States Geological Survey, and Thomas G. Goonan. 2005. "Flows of Selected Materials Associated with World Copper Smelting." Reston, VA: United States Geological Survey. https://pubs.usgs.gov/of/2004/1395/.

This report from the United States Geological Survey (USGS) provides a comprehensive overview of global material flows associated with copper mining and smelting. According to the USGS, the electronics sector is the second largest consumer of copper products after the building and construction industry. Among the important themes of this report is a specific focus on volumes of waste generated by copper mining and copper smelting globally as well as at specific facilities in Chile, Europe, Japan, and the United States. The figures on waste arising in copper mining and smelting help put into perspective the relative magnitude of pollution and waste arising in the mining sector relative to that arising in manufacturing of electronics, and to that arising as post-consumer discards of electronics. Figures available in this report show that annual waste arising at specific individual copper mines and smelters substantially exceed measures of waste arising from post-consumer discarded electronics in whole countries or regions. Copper receives much less attention than that directed toward 3TG minerals. Yet, the electronics industry is far more consequential with respect to the volumes of waste and pollution that can be attributed to mining and smelting for copper than for 3TG minerals. In this respect, this USGS report provides an important demonstration of thinking carefully about scalar mismatch. Solutions to pollution and waste problems need to carefully consider the degree to which they are or are not aligned with overall magnitude of the pollution and waste problems they claim to solve.

Vogel, Christoph, and Timothy Raeymaekers. 2016. "Terr(it)or(ies) of Peace? The Congolese Mining Frontier and the Fight against 'Conflict Minerals.'" *Antipode* 48, no. 4: 1102–21. https://doi.org/10.1111/anti.12236.

This article by geographers Christoph Vogel and Timothy Raeymaekers is an important intervention into broader debates about conflict minerals, including their incorporation into electronic devices specifically. The authors draw on fifteen years' worth of ethnographic fieldwork in eastern Democratic Republic of the Congo (DRC). They are concerned that the prevailing narratives about conflict minerals and the violence and environmental degradation that they are associated with frame the problem in ways that have substantial negative, albeit unintended, consequences for people with the least power in the region. A key issue, the authors argue, is that various initiatives to assure responsible mineral supply chains, such as ITSCI, or legislation such as Dodd-Frank, end up placing an entire economic sector worked by people in highly vulnerable conditions not of their own making under a general climate of suspicion that ends up further marginalizing the very people and places that such initiatives claim to protect. For example, the authors demonstrate top-down decisions made outside the region meant to improve the governance of supply chains originating there quickly led to substantial negative consequences for the estimated 500,000 artisanal miners and their 6–8 million dependents. These negative outcomes include a radical loss of income to people already living in conditions of poverty, an increase in forced migration, the militarization of mines under the DRC's army, and an uptick in recruitment to various militias in the region.

A key takeaway of this research by Vogel and Raeymaekers is that it is crucial to pay attention to the uneven power relations built into moral concerns over the living, working, and environmental conditions of artisanal miners. Governance

214   *Electronic Waste*

mechanisms intended to ban sourcing of conflict minerals from specific regions can substantively enhance the same negative conditions they ostensibly are promulgated to alleviate. That such interventions can have negative unintended consequences is not an argument that advocacy and policy intervention cannot or should not be done. Instead, as with other facets of pollution and waste arising from electronics, it is crucial for proponents of solutions to consider power imbalances and how proposed solutions do or do not appropriately support people and places in subordinate positions relative to those who are already dominant.

Manufacturing

Belkhir, Lotfi, and Ahmed Elmeligi. 2018. "Assessing ICT Global Emissions Footprint: Trends to 2040 & Recommendations." *Journal of Cleaner Production* 177 (March): 448–63. https://doi.org/10.1016/j.jclepro.2017.12.239.

Belkhir and Elmeligi's work is a quantitative analysis of greenhouse gas (GHG) emissions associated with information and communication technology production and use globally. Their study addresses common consumer computing devices such as desktops and notebooks as well as emissions from data centers and communication networks. The authors use lifecycle assessment techniques to estimate energy and emissions associated with the production phase, the use phase, and the overall useful life of various electronics. This article is a particularly helpful summary of calculations of GHG emissions for six categories of common consumer electronics (desktop computers, notebook computers CRT displays, LCD displays, tablets, and smartphones). GHGs are only one class of pollutant associated with the manufacturing and use of electronics. However, data available in Table 9 make it clear that the production phase of electronics is the largest contributor to their overall GHG footprint by far. As such, these data are a good starting point for researchers thinking critically about proposed solutions to the overall pollution and waste problems associated with electronics. No amount of post-consumer recycling will be able to recoup the GHG emissions from manufacturing which are released before consumers have purchased the devices they use and may subsequently discard. Belkhir and Elmeligi's analysis in this respect is not an argument against post-consumer recycling. Instead, it is an empirical demonstration of why post-consumer recycling on its own cannot be a solution to the overall waste and pollution problems associated with electronics.

Ensmenger, Nathan. 2018. "The Environmental History of Computing." *Technology and Culture* 59, no. 4: S7–33. https://doi.org/10.1353/tech.2018.0148.

Environmental history is a sub-discipline of the broader field of history and pays close attention to ecological factors and human action mutually shape one another. Ensmenger's "The Environmental History of Computing" traces these interactive relations between physical geography and history of computing infrastructure over time. The paper focuses on the history and geography of subsea fiber optic cable networks, how the history of railway networks in the continental United States subsequently influenced the physical geography of key portions of internet infrastructure as well as the location of bitcoin operations relative to power plants and electrical distribution grids in the United States. An important part of Ensmenger's historical analysis is that despite being based on seemingly weightless and immaterial components such as bits, bites, and light computation is an inherently physical activity and has both inherently physical environmental resource needs and consequences. This paper is a good demonstration of why it matters to think broadly about the environmental impacts of electronics as occurring throughout their existence rather than suddenly emerging when consumers discard their devices.

Heppler, Jason A. 2017. "Green Dreams, Toxic Legacies: Toward a Digital Political Ecology of Silicon Valley." *International Journal of Humanities and Arts Computing* 11, no. 1: 68–85. https://doi.org/10.3366/ijhac.2017.0179.

Heppler's article is a geographical and historical examination of the transformation of what would become the Silicon Valley region from a predominantly agricultural economy to one now synonymous with "big tech." Although focused on the region as a case study, Heppler's analysis offers broader insight into both environmental and social change wrought by the electronics industry. The article is notable for its creative use of both qualitative historical methods, such as archival research, as well as the use of geographic information systems to track economic, environmental, and social change in the region. By bringing these techniques together, Heppler can critically assess the association of the electronics industry in Silicon Valley with ideas of "clean" and "light" industry, supposedly very different from the heavy industry of the

US industrial Northeast. Among the paper's findings is that sites of substantial industrial pollution associated with electronics manufacturing in the Silicon Valley region tend to cluster in census tracts with higher proportions of people who are lower income and racialized as non-white. In this respect this paper is a good example of how questions of pollution and waste associated with electronics are also bound up with questions of environmental, economic, and social justice. Moreover, with its emphasis on pollution from manufacturing, this paper offers more evidence as to why it is important to consider a broader frame for waste and pollution from electronics than as something that happens only after consumers get rid of their devices.

Lécuyer, Christophe. 2017. "From Clean Rooms to Dirty Water: Labor, Semiconductor Firms, and the Struggle over Pollution and Workplace Hazards in Silicon Valley." *Information & Culture* 52, no. 3: 304–33. https://doi.org/10.7560/IC52302.

"From Clean Rooms to Dirty Water" is a history of labor activism in the Silicon Valley region from the 1950s to the 1980s. The author argues that although this labor activism was principally directed toward enhancing unionization at electronics firms, its success in this respect was limited. On the other hand, this work of labor organizers did have important positive impacts in terms of enhancing the occupational health and safety conditions at electronics manufacturing firms in the region. It also led to broader environmental remediation efforts to mitigate the contamination of the region's drinking water by toxicants released from the manufacturing of electronics in the region.

The article offers important historical understanding of the organizing work undertaken by the Santa Clara Center for Occupational Safety and Health (SCCOSH), the Silicon Valley Toxics Coalition (SVTC), and their allies against the Semiconductor Industry Association (SIA) and major electronics manufacturers. One important facet of this article is the attention it pays to the occupational health and safety impacts of specific chemical toxicants for workers within manufacturing plants as well as their broader environmental consequences. Lécuyer's article is a good example of research into the pollution and waste implications of electronics that pays attention to chemical toxicants whereas a good deal of other research focuses almost exclusively on greenhouse gas (GHG) emissions. GHGs are, of course, important, but an exclusive focus on them can inadvertently erase the other important environmental impacts of the manufacturing of electronics.

Williams, Eric, R. U. Ayres, and M. Heller. 2002. "The 1.7 Kilogram Microchip: Energy and Material Use in the Production of Semiconductor Devices." *Environmental Science & Technology* 36, no. 24: 5504–10.

"The 1.7 Kilogram Microchip" is an important early paper that quantifies the embodied energy and materials in the making of electronics components, particularly semiconductors. Part of the paper's importance arises from its interrogation of the assumption that digitalization leads to dematerialization, that is, that electronics and the transfer of information will lead to a reduction in the material intensity of economic activity. For example, the authors demonstrate quantitatively that the manufacturing of a 2 g dynamic random access memory (DRAM) chip requires 1600 g of fossil fuels and 72 g of chemical inputs. In addition, 32,000 g of water and 700 g of specialized gases (such as dinitrogen, or N_2, which is also a potent greenhouse gas) are also required per chip manufactured. Although the energy and material efficiency of semiconductor manufacturing processes have improved since the publication of this paper, the broader lesson remains true: manufacturing electronics is a material- and energy-intensive process with concomitant negative environmental impacts arising from pollution and waste attributable to the manufacture of devices. When e-waste is construed as only what arises from post-consumer discard of devices, the pollution and waste arising in manufacturing of those devices are missed.

Yu, Jinglei, Eric Williams, and Meiting Ju. 2010. "Analysis of Material and Energy Consumption of Mobile Phones in China." *Energy Policy* 38, no. 8: 4135–41. https://doi.org/10.1016/j.enpol.2010.03.041.

This research is a quantitative analysis of material and energy consumption associated with making and using mobile phones in China. Although the research is focused on a single class of consumer electronic device in a single country, it nevertheless offers important broader lessons about the relationship between solutions proposed to mitigate or eliminate pollution and waste associated with electronics. The paper provides a breakdown of energy consumption in six segments of a phone's lifecycle from raw material production, component manufacturing, mobile phone assembly, packaging and transport, usage, and network operation. The researchers demonstrate that from an energy consumption viewpoint, component manufacturing is by far the largest segment of energy consumption in the overall lifecycle of a mobile phone. Fifty percent of the total energy consumption over the phone's lifecycle is attributable to manufacturing alone. Meanwhile, use accounts for 20 percent. The remainder (30 percent) is attributable to the initial packaging and transport of the device and the energy requirements of the networks to which mobile phones connect to transfer voice and other data. The results of this paper are a good demonstration that postconsumer waste management practices such as recycling cannot on their own mitigate the energy expenditures and associated pollution of activities such as manufacturing, packaging, and transport given that they occur before consumers purchase their devices.

Use and Reuse

André, Hampus, Maria Ljunggren Söderman, and Anders Nordelöf. 2019.
"Resource and Environmental Impacts of Using Second-Hand Laptop Computers: A Case Study of Commercial Reuse." *Waste Management* 88: 268–79.

This article reports the results of a lifecycle assessment (LCA) of second-hand laptop reuse facilitated by an actual commercial reuse business based in Sweden. The study examines the impact of reuse with respect to climate change, human toxicity, and metal resource use. The research considers the transportation of the reused laptops to their major markets, which include destinations in Europe and Asia. Very substantial environmental benefits are found to accrue from reuse of used laptops, even with the significant transport distances required to get the laptops from the reuse business in Sweden to their customers elsewhere in Europe and Asia.

The researchers break out their LCA analysis by individual laptop component as well as providing an overall measure of net impact of reuse. From a climate change perspective, the researchers find that second-hand laptops offer a net benefit of approximately 40 percent compared to new laptops. The study notes that human toxicity exposure occurs predominantly during the primary production of metals for laptops. Thus, those second-hand laptops that are diverted to proper recycling led to a reduction of 75 percent of human toxicity exposure relative to the primary production of metals needed for new laptops. As with other studies, this article clearly shows the predominant negative environmental impacts of electronics occur in the mining of the metals for them and the subsequent manufacture of devices. This research is a good example of the importance of

Resources

reuse of electronic devices as an energy and material conservation strategy that also significantly reduces associated pollution and waste.

Masanet, Eric, Arman Shehabi, Nuoa Lei, Sarah Smith, and Jonathan Koomey. 2020. "Recalibrating Global Data Center Energy-Use Estimates." *Science* 367, no. 6481: 984–6. https://doi.org/10.1126/science.aba3758.

Data centers are rooms or buildings dedicated to housing telecommunications infrastructure such as servers that store and transmit data to users elsewhere. They are a crucial piece of electronics infrastructure that facilitates the transfer of information between geographically dispersed users and data. There is substantial concern about the growing energy demand associated with the growth of data centers globally. The authors of this article offer a critical analysis of energy consumption trends for data centers and suggest that previous analyses overestimate the magnitude and rapidity of the growth in energy use associated with such infrastructure.

The authors combine an analysis of the kinds of services data centers are devoted to along with a variety of changes in the underlying equipment related to energy efficiency to perform their analysis. They note that earlier analyses performed a decade earlier indicated that data center energy use would grow to as much as 1.5 percent of total global electricity used between 2005 and 2010 if trends prevalent at the time continued. Using newly available data that also account for improvements in energy efficiency with real equipment, these authors argue that correlations between the growth in data centers and their overall energy use have been substantially reduced to the point of being effectively decoupled from one another.

Energy consumption is, of course, reliant on energy generation. Energy generation has its own environmental impacts ranging from the ongoing climate emergency when fossil fuels are used for that generation, to the drowning of landscapes for hydro power, to the continued mining for metals required for various renewable energy infrastructure. This makes the link between energy generation for data center operations and specific pollution and waste consequences a complex issue. However, it is, indeed, a significant finding if a decoupling between the growth of data centers and their overall energy consumption has in fact occurred.

As important as the findings of this paper are, there is a note of caution offered by the authors that is quite telling. Having found the relationship between data center growth and energy consumption growth decoupled from one another, the authors point out that there is no guarantee this relationship will continue. They argue for a wise use of time to prepare for possible increases in demand for energy for data centers. Published in 2020, before the wide-scale deployment and adoption of large language learning models (LLMs or "AI"), the authors' note of caution is likely prescient as these newer compute applications are, to date, highly energy intensive. It may be that the decoupling measured by these authors is only temporary and that the data center–energy consumption relationship ends up conforming to the Jevons Paradox or rebound effect.

Parajuly, Keshav, James Green, Jessika Richter, Michael Johnson, Jana Rückschloss, Jef Peeters, Ruediger Kuehr, and Colin Fitzpatrick. 2024. "Product Repair in a Circular Economy: Exploring Public Repair Behavior from a Systems Perspective." *Journal of Industrial Ecology* 28, no. 1: 74–86. https://doi.org/10.1111/jiec.13451.

The authors of "Product Repair in a Circular Economy" report the results of a large international survey of people participating in public events for the repair of electrical and electronic products, with 922 respondents in fourteen countries. Results were grouped into three categories of factors that influence the peoples' intention to repair: consumer behavior factors such as attitudes and beliefs, technical and economic settings, and policy interventions that could support repair.

The results of the study show that multiple factors play a role in shaping people's intention to repair. One of the strongest factors influencing intent to repair is peoples' attitudes and values along with respondents' perception of their skills or abilities to affect repair. However, an important finding of the study is that individual attitudes, norms, and values are insufficient on their own to sustain and enhance intention to repair. The authors note the importance of technical, economic, and policy infrastructure that conditions people's intentions to repair. Survey responses show a statistically significant difference between perceptions about the degree of difficulty of repairing electrical appliances versus electronic equipment. Respondents' perceptions of electronic equipment are that this class of products is considerably more difficult to repair than are appliances. Ultimately, the authors argue that while education or awareness can influence peoples' attitudes and behaviors about repair in positive ways, changes in personal norms are insufficient on their own to enhance repair. The authors recommend device manufacturers change design practices to enhance repairability by non-experts. As such, the authors recommend policy changes that would require design for repair as well as changes to warranty regulations that do not allow companies to restrict third-party independent repair.

Post-Consumer Discard

Althaf, Shahana, Callie W. Babbitt, and Roger Chen. 2019. "Forecasting Electronic Waste Flows for Effective Circular Economy Planning." *Resources, Conservation and Recycling* 151 (December): 104362. https://doi.org/10.1016/j.resconrec.2019.05.038.

The authors of this paper offer a predictive material flow analysis (MFA) that can be used for forecasting purposes in reorganizing economic activity away from a linear to a circular orientation. MFA is a general approach to modeling how flows and stocks of materials move through a given system. The authors apply this general technique to the throughput of materials of various categories of consumer electronics in the context of the United States over time and offer a short-term forecast of changes out to the year 2025 (i.e., six years after the publication of the research). Their modeling suggests that mature electronic products such as printers, desktops, CRT monitors, laptops, LCDs, and other monitor types have already reached their peak sales volume in the United States. Their results point broad scales shift in the generation of post-consumer e-waste in the United States between 2010 and 2025. As depicted in this paper, household e-waste in the United States likely peaked in 2016 at approximately 2.3 million metric tons and has been in decline since then. The authors forecast that by 2025 the amount of post-consumer e-waste arising in US households will fall to approximately 600,000 metric tons. What is more is that the overall composition of household e-waste arising will change shifting from a predominance of CRT monitors to a predominance of flat panel televisions and monitors, tablets, laptops, and desktops. These changes are important because they change the material composition—and thus the economics—of stocks of e-waste that flow to waste management operations such as recycling and disposal.

MFA analyses are highly quantitative and technical. However, non-technical readers of this research will find much of use in it. For example, the authors can demonstrate to key trends in the generation of end-of-life electronic devices in the United States over time. First, the authors can document clear trends in the reduction of product cycles between the 1960s in the early 2000s. Their findings indicate that the time between when a product first enters the market and when sales subsequently decline has shortened from almost forty years to less than ten. What this shortening of product lifecycles represents is a fourfold increase in the rate of throughput of the materials and energy they embody. A second trend is also important and conditions the consequences associated with these shortened product lifecycles. The authors demonstrate a trend toward lighter and lighter weight products over time. In other words, because less and lighter materials are being used in products there is a decline in the throughput of material mass embodied in those products. This trend toward lighter weight devices has important implications for policies directed toward managing end-of-life electronics such as through recycling. Many existing electronics recycling programs rely on weight metrics to determine funding formulas to support those programs. The results of this study point to an emerging challenge for such e-waste recycling systems. Somewhat ironically, the systems are handling a larger number of devices, but a declining total mass of devices due to light weighting. This uncoupling between the number of devices handled and the mass of total devices managed can lead to unintended negative consequences for sustainable funding of electronics recycling systems. Part of the value of the research reported in this paper derives from its predictive results for both mature devices (e.g., monitors, laptops, desktops) and emerging electronic products (e.g., fitness trackers, drones). The forecasted shifts in the flows and stocks of materials embodied in end-of-life electronic devices are important information for policymakers seeking to maintain viable post-consumer recycling programs for electronics. Based on their results, the authors recommend policymakers to devote more attention toward regulations that will encourage brand manufacturers to increase the durability, repairability, and reusability of their products.

Ceballos, Diana Maria, and Zhao Dong. 2016. "The Formal Electronic Recycling Industry: Challenges and Opportunities in Occupational and Environmental Health Research." *Environment International* 95: 157–66. https://doi.org/10.1016/j.envint.2016.07.010.

This paper reports results from a systematic review of occupational and environmental health hazards of formal e-waste recycling facilities. The authors note that considerable attention has been paid to occupational and environmental health hazards of informal e-waste recycling in a variety of locations. They seek to investigate whether an implicit assumption that formalization is a solution to the occupational and environmental health hazards of informal recycling.

The studies reviewed by these authors come from multiple countries including China, Canada, Finland, France, the United States, Sweden, and Thailand. This review found that while formalization can reduce occupational and environmental health risk exposure, problematic exposure of workers to heavy metals and toxic chemicals still occurs even in formal recycling facilities operating under high occupational health and safety standards. Studies reviewed showed that workers in such facilities are often exposed to toxicants above-recommended thresholds. Moreover, in one study of an e-waste recycling facility in the United States, children of workers were found to be exposed to toxicants brought home on the work clothing of parents employed at the facility.

This review of occupational and environmental health risks associated with formal electronic recycling is important given the substantial attention to the informal recycling of electronics, especially at sites in so-called developing countries. Occupational health and environmental conditions in those informal settings are, of course, concerning. What the results of this paper show, however, is that while those risks can be reduced through formalization, this does not happen automatically. Moreover, even facilities that operate in well-regulated jurisdictions still experience conditions that expose workers to occupational and environmental health hazards that exceed mandated threshold. An implication of this research is that formal electronic waste recycling is, on its own, an insufficient solution for reducing occupational health and environmental exposure to toxicants. The results of this paper suggest why it is important to direct solutions upstream before electronic products are manufactured in the first place if the goal is to reduce or eliminate toxicants and the pollution and waste they are associated with.

Puckett, Jim, Leslie Byster, Sarah Westervelt, Richard Gutierrez, Sheia Davis, Asma Hussain, and Madhumitta Dutta. 2002. "Exporting Harm—The High-Tech Trashing of Asia." Seattle: Basel Action Network Silicon Valley Toxics Coalition. http://olo.ban.org/E-waste/technotrashfinalcomp.pdf.

"Exporting Harm" is an important investigative report by the environmental NGO, Basel Action Network. This report is one of the most frequently cited documents in the research literature on e-waste when it is framed as an end-of-pipe, post-consumer problem. This report is the source of some of the most frequently cited statistics about the quantity of post-consumer e-waste exported from the United States. The report uses a mixture of methods that includes fieldwork in China, India, and Pakistan as well as interviews and photography to document the problem.

Importantly, even though this report has become an important touchstone in the discussion of e-waste as an end-of-pipe problem, the authors note that the most impactful solutions are to be found upstream of managing end-of-life devices. For example, the report recommends substituting non- or less toxic materials in manufacturing and in the final design of devices. It also recommends changes to design practices that enhance durability and possibilities for device upgrading as well as repair and reuse. The latter factors are important design interventions that can lengthen the useful life of devices, acting as a conservation mechanism for the energy and materials embodied in them. The report also recommends implementing genuine extended producer responsibility (EPR) regulations that would force manufacturers to pay for end-of-life device management, thus incentivizing them to make less toxic, longer lasting, and more easily manageable devices. While this report has met with some criticism and controversy for the reliability of the statistics it reports, it is nevertheless important for its early recognition of the role of upstream solutions to overall problems of pollution and waste arising from electronics.

Journals

ACS Sustainable Chemistry & Engineering

ACS Sustainable Chemistry & Engineering is an American Chemical Society (ACS) journal of peer-reviewed research that covers green(er) chemistry in manufacturing and engineering, the revalorization of wastes as resources, alternative energy, and lifecycle assessments, among other topics. The journal publishes a variety of article types including letters that convey preliminary research of broad significance, full research articles, viewpoints which offer opinions, and perspectives that identify research gaps and document debates and disagreements amongst researchers within given fields or topics covered by the journal. Research published in this journal tackles problems associated with pollution and waste arising in the mining for, the manufacturing of, electronics as well as their discard, management, and processing.

Chemosphere

Chemosphere is a peer-reviewed journal of environmental chemistry research. Articles published in the journal cover the identification of chemicals in the environment, their quantification, how they behave within environmental media, toxicology, treatment, and remediation. As a journal focused on environmental chemistry, research published in it covers toxins (those chemicals that arise in natural environments) and toxicants (chemicals that arise from manufacturing and industrial processes that would not otherwise be found in nature). Articles covering toxicology and risk assessment include research on adverse effects of chemicals in different environmental media such as atmospheric, terrestrial, and water-based systems, epidemiological studies, and biomonitoring. Research published in the journal on treatment and remediation focuses on technological means to eliminate or mitigate environmental contamination. Articles published in *Chemosphere* cover a wide range of issues associated with electronics, including pollution and waste arising from associated mining, manufacturing, use, reuse, and discard.

Environmental Pollution

Environmental Pollution focuses on peer-reviewed research investigating the effects of pollution on ecosystems and human health. Articles in the journal focus on a wide range of issues. These include identifying sources and occurrence of pollutants in different environmental media (e.g., atmosphere, hydrosphere), food, and organisms including human bodies. The journal also publishes research on emergent contaminants of concern. These are categories of environmental contamination that are novel. Among the examples of such emergent concerns highlighted by the journal is electronic waste. The latter generally refers to contamination arising from processing end-of-life, post-consumer electronics. However, many articles published in the journal also cover pollution and waste issues associated with the mining for, and manufacturing of, electronic devices.

Environmental Science and Pollution Research

Environmental Science and Pollution Research is a peer-reviewed journal of environmental sciences. It is focused on research that covers, among other things, biological and ecological systems in various spheres of the Earth system (e.g., atmosphere, hydrosphere, biosphere), remediation and restoration of ecosystems, environmental monitoring, pollutant risks, environmental economics, and natural hazards. The journal publishes research articles as well as news, views, commentary, and editorials covering the broad spectrum of topics in its remit. Research on issues of pollution and waste associated with the mining for, the manufacturing of, as well as the discard of electronic devices can be found in the journal.

Environmental Science and Technology

Environmental Science and Technology is a peer-reviewed journal covering a broad range of topics investigated by environmental scientists and engineers. It aims to publish research that is relevant to diverse audiences, including scientists, but also policymakers and broader publics gathered around environmental concerns. As such, the journal publishes research articles, replies and rebuttals, viewpoints (i.e., opinion pieces), reviews of subfields and specific topics, and policy analysis among other article types. Articles published in the journal offer a broad range of analyses on pollution and waste issues associated with electronics, including mining, manufacturing, and post-consumer discard, management, and processing.

Journal of Cleaner Production

The *Journal of Cleaner Production* is a peer-reviewed journal focused on research geared toward the prevention of waste. It covers cleaner production techniques from technical perspectives as well as assessments of environmental impacts of current production methods, analysis of policy and legislation, and assessment of corporate sustainability and responsibility, among other topics. The journal is a good source of research and analysis of pollution and waste arising upstream in the manufacturing segments of the lifecycles of electronics and that arising from post-consumer discards of electronics.

Journal of Hazardous Materials

The *Journal of Hazardous Materials* is a peer-reviewed publication that publishes research under the broad umbrellas of environmental sciences and engineering. The journal focuses on contaminants that are or could be encountered in realistic environmental conditions. Papers published in the journal include both field- and laboratory-based research. Research published in the journal includes analysis of how different contaminants move through and accumulate in different environmental media, such as trophic levels, the atmosphere and hydrology, among others. Papers published in the journal cover the release and ecological behavior of hazardous materials arising from mining, manufacturing, and post-consumer electronic discards.

Resources, Conservation, and Recycling

Resources, Conservation, and Recycling is a peer-reviewed journal focused on the sustainable use and conservation of resources. A key focus of the journal is research that maps pathways toward more sustainable production and consumption systems. The journal emphasizes research geared toward improving technical processes, economic and business models, and policy that contribute to more efficient use of resources and/or the reuse of materials derived from waste. Papers published in the journal cover a wide range of geographic scales from the regional, to the national, to the international as well as whole economic sectors and/or their sub-systems. The journal specifically excludes purely laboratory-based research except for cases that have direct, practical applications outside the laboratory. Research in the journal includes research on post-consumer e-waste as well as analyses of pollution and waste arising in manufacturing.

Science of the Total Environment

Science of the Total Environment is a peer-reviewed journal focused on research in the natural/physical sciences. Among other topics, research published in the journal covers ecotoxicology, the environmental impacts of industrial processes and economic sectors, and environmental remediation. The journal publishes articles assessing the toxicological hazards associated with downstream electronic waste processing. It also publishes research related to pollution and waste arising in the manufacturing of electronics.

Waste Management

Waste Management is a peer-reviewed journal focused on various issues related to solid waste management. Topics covered by the journal include studies of generation and characterization (e.g., material composition) of solid waste, waste minimization, recycling and reuse as well as economic, policy, and regulatory analysis. Research published in the journal often uses quantitative lifecycle assessment and economic assessment methods. Notably, the journal focuses on municipal, agricultural, and certain special wastes such as hazardous household as well as hazardous and non-hazardous industrial solid waste. However, the journal specifically excludes research on waste arising in the mining and metallurgical processing (e.g., smelting) sectors.

Waste Management & Research: The Journal for a Sustainable Circular Economy

Waste Management & Research is a peer-reviewed journal devoted to scientific research on waste management accessible to a diverse audience of academics, policymakers, industry analysts, managers, planners, and public health practitioners. Research in the journal focuses primarily on solid waste but does so throughout the lifecycle of products and materials. Thus, research published in the journal can be found on solid waste issues associated with mining, manufacturing, use, and discard of electronics.

7

Chronology

What follows is a chronological list of key moments in the emergence of the issue of pollution and waste arising from electronics as being of public concern. The chronology offered is not exhaustive. Instead, it is intended to demonstrate that contemporary concerns about e-waste understood as an end-of-pipe, post-consumer waste management problem is a very particular framing of the issues situated within broader concerns about the environmental consequences of the electronics industry.

1972   The US Department of the Interior, Bureau of Mines publishes a report called, "Recovery of Precious Metal from Electronic Scrap." The report identifies discarded electronics from military and civilian applications as substantial sources of potentially recoverable precious metals. It is exactly this kind of end-of-life electronics which some thirty years later would come to be identified as a novel and rapidly growing waste management problem called "e-waste." The conceptual and practical shift from treating discarded electronics as scrap to treating them as waste is a good example of the malleability of "wastes" and "resources."

1976   Santa Clara Center for Occupational Safety and Health (SCCOSH) is formed in San Jose, California. SCCOSH activists and researchers document the occupational health and safety risks associated with the use of chemical toxicants in the electronics manufacturing facilities of the region. The group also plays an important role in researching the environmental effects of the releases of toxicants into the broader environment, including groundwater, in the region.

A data center for a bitcoin and crypto mining farm. Although there is some evidence that total energy use is declining relative to the increase in compute time at data centers, new applications such as machine learning (sometimes called artificial intelligence) may soon result in highly energy-intensive data centers. (Michal Bednarek/Dreamstime.com)

232   *Electronic Waste*

1977 The Pacific Studies Center, a nonprofit public interest research center, publishes a report called, "Silicon Valley: Paradise or Paradox?: The Impact of High Technology Industry on Santa Clara County." The report identifies many problematic social, economic, and environmental issues associated with electronics manufacturing in the region. It notes the negative effect on housing affordability, the preponderance of racialized, non-unionized workers and their precarity, and the release of toxicants such as trichloroethylene (TCE) in the workplace and the broader environment. The report will prove to be a prescient warning about the negative social, economic, and environmental consequences of electronics manufacturing in Silicon Valley and beyond.

1980 The US National Institute of Occupational Safety and Health publishes a report on significant exposures of workers to chemical toxicants at an electronics manufacturing facility in Sunnyvale, California. Such exposures represent the movement of materials with toxic properties into human bodies, rather than the electronic devices being manufactured. As such, this is an early example of a broader understanding of pollution and waste arising from electronics than what the overly narrow framing of e-waste as an end-of-pipe problem that would emerge in the 2000s would consider.

1981 Contamination of drinking water by leaking underground chemical storage tanks at electronic manufacturing facilities in San Jose, California, is confirmed.

1982 The Silicon Valley Toxics Coalition (SVTC) forms out of a series of meetings organized by SCCOSH and its supporters. SVTC is devoted to protecting residents from toxic exposures in the region, largely resulting from mismanagement of chemical toxicants in the electronics manufacturing industry. The work of SVTC would go on to be fundamental to the later designation by the US Environmental Protection Agency of twenty-nine "Superfund" sites related to electronics manufacturing facilities in the region.

1983 SVTC's advocacy leads to the passage of the first right-to-know laws and hazardous materials handling ordinances in the Silicon Valley region. These regulations would go on to become part of key federal

Chronology

233

regulations related to Superfund legislation and the creation of the United States' Toxic Release Inventory (TRI) database. TRI would go on to become a publicly accessible database of pollutant releases from all industrial sectors operating in the United States.

1985   IBM corporate headquarters is alerted by one of its own chemists to a cancer cluster at an IBM facility in San Jose, California. IBM commissions a study based on mortality records for all of its employees. The study finds suggestive links between specific job categories, chemical toxicants, and increased risk of various kinds of cancer. In the same year, a California Health Department study finds higher than expected rates of miscarriages and birth defects in areas adjacent to leaking underground chemical storage tanks at Fairchild Semiconductor and IBM facilities in Silicon Valley. The California Health Department report is early confirmation of broader, negative environmental impacts of toxicants from the industry.

1987   The governing body of the United Nations Environmental Program begins negotiations on a multinational agreement to regulate the international movement of hazardous wastes. These negotiations would eventually lead to the Basel Convention on the Transboundary Movement of Hazardous Wastes and Their Disposal. The Basel Convention includes the regulation of post-consumer discarded electronics.

1988   Significant contamination of groundwater in Kimitsu City near Tokyo, Japan, is publicly revealed. Studies seeking the source of contamination lead back to electronics manufacturing facilities in the region. Trichloroethylene (TCE) is a chemical toxicant of concern found in the water. Research indicates that spills of TCE occurred both on the surface and underground from leaking chemical storage tanks, like the situation in Silicon Valley.

1989   Thousands of people protest outside an IBM electronics manufacturing plant in San Jose, California, on Earth Day. The people marching express their opposition to the more than 1,500,000 pounds of chlorofluorocarbons (CFCs) that the company itself reported in its own Toxic Release Inventory data. CFCs were in use as a solvent in IBM's manufacturing processes, but also played an important role

234 *Electronic Waste*

in the formation of localized air pollution and damaging the global ozone layer. The company found ways to substitute soap and water as an alternative solvent. Also, in 1989 the US EPA designates an area of approximately 25 km² as a zone of interest due to the contamination of soil and groundwater by electronics manufacturing facilities in the Silicon Valley region. The area so designated is dubbed the Middlefield-Ellis-Whisman (MEW) Superfund Study Area, named after the roads that form its borders.

1990 The Silicon Valley Toxics Coalition launches the Campaign for Responsible Technology (CRT, later becoming the International Campaign for Responsible Technology or ICRT). CRT's initial focus is raising awareness about electronic industry's lobbying for what the group calls corporate welfare such as tax breaks and rollbacks of environmental and occupational health and safety regulations. Among the group's concerns are the pollution and waste consequences of electronics manufacturing for the communities in which those manufacturing facilities are located. ICRT expands its work beyond the United States as the globalization of the electronics industry intensifies in the 1990s.

1991 Bamako Convention prohibiting the transboundary shipment of hazardous wastes to African countries is adopted. Although similar to the Basel Convention in some respects, the Bamako Convention includes provisions for cleaner production that aim to eliminate or mitigate pollution and waste before they arise in the first place.

1992 The Basel Convention comes into force. Singapore and South Korea pass some of the earliest legislation covering management of some categories of post-consumer e-waste. In the same year CRT (later, ICRT) convinces the US Congress to expand the remit of SEMATECH, a not-for-profit research and development consortium originally formed in partnership between US government and US-based semiconductor manufacturing companies, to include the development of cleaner production techniques in the industry. Research into cleaner production means pollution and waste may be reduced or eliminated from the second most consequential lifecycle phase of electronics after mining. The board of SEMATECH votes to

Chronology 235

sever its partnership with US government agencies four years later. Meanwhile, results from a study of IBM workers begun five years earlier are published and they show that women exposed to chemical toxicants at IBM semiconductor plants experience disproportionately high miscarriages. The IBM study becomes the basis of industry-wide efforts to phase out some chemical toxicants from their manufacturing processes.

1994 The North American Free Trade Agreement (NAFTA) comes into effect. The international trade agreement has sweeping consequences for the economic organization of the signatory countries of Canada, Mexico, and the United States. A provision of NAFTA creates the Commission for Environmental Cooperation (CEC). Among the remit of the CEC is the collection and publication of pollution release and transfer data of firms operating in the NAFTA region. Reporting of such data to the CEC is initially voluntary but becomes mandatory as of 2006. CEC data track the release of over 600 different chemicals deemed to be toxic hazards across all industries operating in the NAFTA region. These data are publicly available and permit the tracking of toxic pollutant releases from specific economic sectors, such as electronics manufacturing, at the individual facility level anywhere they operate in Canada, Mexico, or the United States.

1997 A fire breaks out at an electronics manufacturing facility in the Hsinchu area of Taiwan. Firefighters responding to the emergency fall ill and spend weeks recovering in hospital from having inhaled fumes associated with the fire. The incident becomes an important wakeup call about the environmental risks of electronics manufacturing in the region. Reports of illegal dumping of toxic solvent waste from electronics manufacturers into the region's groundwater systems also continue to emerge.

1998 The American Electronics Association (AEA) begins lobbying against legislative changes being considered in Europe to phase out toxicants and to require that manufacturers assume full financial responsibility for collecting and processing discarded electronic devices. The AEA convinces US trade representatives that the legislation being contemplated in Europe represents a threat to American business

236   *Electronic Waste*

interests. A year later the United States threatens to seek redress via a dispute tribunal at the World Trade Organization (WTO). In response, European legislators weaken the provisions for phase out of toxicants by lengthening the time over which the requirements will come into effect.

1999 The Silicon Valley Toxics Coalition publishes a report called "Just Say No to E-waste." It profiles the proliferation of discarded electronic equipment at city dumps in the Silicon Valley region. The report's framing of discarded electronics as a waste problem contrasts with the framing of the same type of equipment over twenty years earlier by the US Bureau of Mines as a potential source of precious metals. The malleability of what counts as waste versus what counts as a resource is demonstrated in the contrast of these two reports.

2000 Trade industry magazine *Electronic Business* publishes an article called, "Ready for Recycling?" about an end-of-life electronics recycling program run by HP at a plant in Roseville, California. The article profiles the processing of what it calls, "electronic waste" at the facility and at a smelter in Canada. With the publication of the Silicon Valley Toxics Coalition report of the year before and the publication of BAN's "Exporting Harm" report two years later, the tight framing of e-waste as an end-of-pipe, post-consumer waste management problem is well underway. One of the effects of this tight framing is that after this period it will become increasingly difficult to widen the conversation such that e-waste includes pollution and waste arising during all phases of the lifecycle of electronics.

2002 The European Union promulgates the Waste Electrical and Electronic Equipment (WEEE) directive. The directive uses a combination of voltage and equipment category to define what will count as waste electrical and electronic equipment under the legislation. The EU WEEE directive becomes the first international regulatory framework for managing end-of-life electronics. In the same year, a US-based environmental NGO called the Basel Action Network (BAN) releases a report called, "Exporting Harm: The High-tech Trashing of Asia." The report goes on to generate substantial media attention on e-waste as an end-of-life, post-consumer waste management problem.

Chronology

2003   California becomes the first US state to pass regulations for the collection and processing of post-consumer e-waste. Over the next decade twenty-five more states would pass similar legislation, the most recent covering the District of Columbia and passed in 2014.

2004   Alberta becomes the first province in Canada to pass legislation regulating the collection, management, and processing of post-consumer e-waste in the country.

2007   South Korea passes legislation covering the recycling of electrical and electronic equipment and vehicles. The legislation represents a significant expansion of existing legislation first passed in 1992.

2009   China enacts legislation to regulate domestically generated post-consumer e-waste in the country.

2010   The US EPA amends its 1989 decision regarding the MEW study area to include contamination of ambient air in homes and buildings within the study area resulting from the migration of toxicants from soil and groundwater. Some of the buildings affected include facilities operated by Google as well as people's private homes.

2011   Nigeria passes national legislation governing the collection, management, and processing of post-consumer e-waste. The legislation covers the same categories of equipment as those in the EU WEEE directive, but also contains notable differences. For example, Nigeria's legislation allows for the importation of used electrical and electronic equipment under certain conditions. It also explicitly includes reference to the waste hierarchy with reference to reduce, repair, reuse, recycling, and recovery in that order.

2012   The US Federal Remediation Technologies Roundtable, an inter-agency working group, releases a presentation called, "Managing a Large Dilute Plume Impacted by Matrix Diffusion: MEW Case Study." The presentation reports on a highly technical study of remediation of toxicants released into the environment by electronics manufacturers within the Middlefield-Ellis-Whisman (MEW) Superfund Study Area of Silicon Valley. Among the findings highlighted in the presentation is that remediation techniques will require anywhere between 500 and 700 years of operation before the area reaches contemporary EPA standards for drinking water.

2013 Japan passes legislation covering the collection and recycling of twenty-eight categories of electronic and electrical equipment. The new legislation is an expansion of existing regulations that have been under discussions since the 1990s and came into effect in 2001.

2016 Ghana passes national legislation regulating the collection, management, and processing of post-consumer electronic waste. The legislation covers such a waste arising both domestically in Ghana and due to imports from abroad.

2022 Colorado and New York are the first US states to pass right to repair legislation covering some categories of consumer electronics. Three more states pass similar legislation over the next two years: California, Minnesota, and Oregon. Meanwhile, also in 2022, both the European Union and the United States pass major legislation intended to bolster semiconductor manufacturing in their respective jurisdictions. Both pieces of legislation authorize the transfer of tens of billions of dollars of public money to subsidize semiconductor manufacturing facilities in the two regions. Much of the impetus for enacting the legislation includes American and European fears about growing competition from China at the cutting edge of semiconductor manufacturing.

2024 The European Union adopts new regulations promoting the repair of goods, including electronics. The regulations are adopted in part to support sustainability goals premised on the mitigation or elimination of pollution waste as well as the loss of materials deemed critical to future economic production. Right to repair activists continue their advocacy for improved regulations in the EU arguing that the regulations adopted to date do not go far enough in their support of repair. Meanwhile in the United States, the EPA finalizes rules requiring public drinking water utilities to reduce PFAS pollution below a regulatory threshold. Wastewater from semiconductor manufacturing facilities is a significant source of PFAS pollution. In the same year the US Department of Commerce and the National Institute of Standards and Technology release the first draft environmental assessment of a new semiconductor facility approved under the CHIPS Act.

Glossary

carbon fixation: An undue preoccupation with carbon dioxide and its global warming effects as a singular indicator of overall pollution and waste arising from some industrial process such as the mining for, and manufacturing of, electronics. Many other pollutants besides carbon dioxide contribute to global heating. The effects of industrial pollution and waste also include more than the climate emergency, for example, toxic contamination.

conservation: Occurs in the context of raw material resource management when recycling leads to a reduction in the extraction of raw materials from the environment.

critical minerals: Lists of specific materials deemed to be of strategic economic importance. Such lists vary from jurisdiction to jurisdiction, but typically include many materials relevant to the manufacture of electronic devices. Common examples include metals such as cobalt, indium, and niobium.

decoupling: The separation of material and energy throughput from economic development. In a genuinely decoupled scenario, any increment of economic development does not also entrain a proportional increment of material and energy throughput.

disposability: A characteristic of expendability designed into many contemporary consumer products, including electronics. Expendability may have both positive and negative consequences. For example, the expendability of medical devices implanted in people is an important form of infection control. On the other hand, disposability can be an economic tactic to enhance profit margins by making products that consumers must repeatedly replace.

downstream: A description of the relative location of action after that which precedes it and from which pollution and waste may arise. Manufacturing of electronics is downstream of the mining of materials for that manufacturing.

efficiency: A ratio of useful work performed compared to the total energy and/ or materials used to perform that work. In the context of electronics, energy efficiency—or the amount of electricity required to perform a unit of computation— is a common metric of efficiency.

electronics: A category of devices that operates using microchips, transistors, and related components to direct electrical currents.

end-of-life electronics: A description of electronic devices that are nominally unusable or unwanted. The description is a metaphor based on the notion of the lifecycle of organisms. The description can cover situations that are both relative and absolute. In relative situations, a given user of a device may find it unusable or unwanted for

240   *Glossary*

their purposes but, if someone else gains access to that same device they may find that it suits their needs. In absolute terms, an electronic device may be considered end-of-life if it can no longer perform even its basic functions such as powering on, although even then repair may enable the device to become useable again.

end-of-pipe: A metaphor based on the idea of a physical cylinder typically used to convey fluids such as water or gas. The metaphor usually refers to a moment when one actor in a system relinquishes control of a product to a subsequent actor in the system. In that sense, the metaphor typically refers to post-consumption situations, that is, when individuals or households get rid of products they have purchased and pass control of them to waste management actors (e.g., a public or private waste collection organization).

environmental justice: The idea and practice of fairness and equity with respect to peoples' right to live in healthy ecological conditions that are free from harms associated with toxic pollution.

environmental racism: The intentional or unintentional disproportionate exposure to toxic harms of people singled out by arbitrary groupings of phenotypic (e.g., skin color) and/or genotypic (e.g., ability to digest milk) characteristics that do not systematically co-vary.

extended producer responsibility: A policy strategy for financing the management of pollution and waste arising from a given category of manufactured products that sources its funding from the original manufacturers of those products.

flow: A metaphor used to describe the movement of materials and/or energy through a system (such as an industrial system that manufactures electronics). The movement of toxicants through ground water or a soil matrix into the atmosphere would be an example of a pollutant flow.

flow control: A metaphor used in waste management scenarios to describe the governance of the movement of solids, liquids and/or gases deemed to be subject to waste management regulations of some kind.

form factor: A set of metrics or specifications that define the physical size and/or shape of a device and/or its sub-components.

hazardous waste: A categorization of particular solids, liquids, and gases deemed to be waste by a regulatory authority that must be handled with special care. Solids, liquids, and gases that are deemed hazardous typically exhibit characteristics such as toxicity and/or reactivity (e.g., chemical or radiological).

incommensurability: The inability to find common standards or criteria by which to assess the status of different things. In a waste management context incommensurability occurs when, for example, metrics of mass versus toxicity are in play. Materials deemed to be waste may have the same mass but differ wildly in terms of toxicity.

indeterminacy: A description of a system or process, the full characteristics or behavior of which, cannot be precisely agreed upon or known.

Glossary

leachate: Water that moves through materials such as those in a landfill and, in so doing, entrains with it some of the characteristics of the materials through which it moves. In a waste management setting, leachate is often used to describe the movement of water through a landfill that picks up some or all of the chemical and biological characteristics of the materials through which it moves in that landfill.

lifecycle: A metaphor based on the idea of the birth, life, and death of an organism. The lifecycle metaphor is often applied to manufactured products, such as electronics, and is meant to describe all of the phases of such products from the mining of raw materials for them, to their manufacture, distribution, use, and eventual discard.

lightweighting: A process by which manufacturers find ways to use fewer and/or smaller masses of materials to make their products.

magnitude: The size or extent of something.

modern waste: A term used to describe contemporary solids, liquids, and gases deemed to be waste and which tend to have characteristics of being synthetic, behaving in ways that are unpredictable, and being heterogeneous.

non-governmental organization: A group of people and a collection of infrastructure gathered for a particular purpose, but outside the formal realms of public administration. In the context of electronic waste, many nongovernmental organizations have formed to advocate for the mitigation and/or elimination of pollution and waste arising from electronics.

planned obsolescence: A combination of design, manufacturing, and marketing strategies that leads to the premature end of a product's useful life to increase sales. There are many critics of planned obsolescence but, there are also commentators who argue that what appears to be the deliberate shortening of product lifecycles is actually just the effect of continual improvement. With respect to electronics, obsolescence can be a result of such strategies applied to hardware, software, or both.

polluter-pays-principle: A proposition that forms the basis of a claim that it is right to assign the economic cost(s) of mitigating or eliminating damage and/or harms arising from the release of harmful substances from a given process. Typically, the principle is used to refer to the responsibilities of upstream actors such as mining and manufacturing firms. In practice, however, the polluter-pays-principle is also used to assign costs to consumers of particular categories of manufactured products such as electronics.

pollution: The release of ecologically damaging substances into environmental media such as air, land, and/or water.

pollution release and transfer registry: A database of emissions, movements, and dispositions of ecologically damaging substances from primary (i.e., raw material extraction) and/or secondary (i.e., manufacturing) industries. Pollution release and transfer registries may provide public access to the data they contain.

post-consumer: Usually used to describe a stage in the lifecycle of a product after individual people or households are ready to rid themselves of that product.

precautionary principle: A proposition widely adopted in various fields such as law and engineering that advocates for measures that may prevent unanticipated dangerous situations from arising and/or mitigate their negative consequences if and when they do occur. A legal example of the precautionary principle would be a regulation that demands chemical manufacturers to demonstrate that new chemicals are not ecologically harmful before those chemicals are allowed to be used in industry.

preservation: Occurs in the context of raw material resource management when recycling leads to the maintenance of ecosystem complexity and functioning.

rebound effect: Increases in aggregate energy and materials throughput in an industrial system that result from earlier improvements in efficiency of energy and material use per unit of work. Improvements in efficiency initially lead to less energy and material throughputs but, these improvements diminish or disappear overtime because of cost savings induced through initial efficiency improvements enabling more energy and materials to be used overall. The term is also known as the Jevons Paradox, gaining its name from William Stanley Jevons who was a nineteenth-century political economist studying resource use and efficiency.

recycling: A wide variety of processes that transform materials and/or energy that would otherwise be discarded from a system back into feedstock for the manufacture of new products and/or new rounds of energy use. Recycling is widely associated with post-consumer waste management practices. However, recycling can occur anywhere in industrial systems from raw material extraction, to manufacturing, to consumption, and beyond. An example of materials recycling would be when a manufacturer directs factory floor shavings of aluminum from machining new product (e.g., a computer housing) back into the manufacturing of additional new product. An example of energy recycling would be the recirculation of air warmed by the operation of computer servers throughout a data center as a form of wintertime thermal comfort for people working in the facility.

resource: Something that is useful. Resources are usually defined in anthropocentric terms as things that are found and/or made to be useful to humans.

scalar mismatch: The misalignment or misspecification of actions deemed to solve a given problem. In the context of problems of pollution and waste, scalar mismatch arises when, for example, post-consumer recycling is put forward as a solution to overall pollution and waste arising from an industrial system.

stock: A metaphor used to describe the accumulation or storage of materials and/or energy in a system (such as an industrial system that manufactures electronics). The accumulation of toxicants in a ground water system would be an example of a stock of pollutants.

sufficiency: A condition of adequacy. In the context of solutions to pollution and waste arising from electronics, sufficiency offers a concept and practice of computer use that slows down product cycles and encourages longer term use of computing devices that already exist rather than the default practice of purchasing new devices.

Glossary 243

Superfund: An environmental remediation program created in 1980 in the United States and administered by the Environmental Protection Agency (EPA). The Superfund program is premised on the polluter pays principle. As such, the EPA engages in work to identify the party or parties responsible for the contamination of a given site, typically a location of historical or ongoing raw material extraction or industrial production. In cases where a party cannot be identified, for example when a company has gone bankrupt or being absorbed by another company in a buyout, the EPA can fund cleanup itself using monies collected from taxes on petroleum and chemical manufacturers.

threshold theory: A proposition that a definitive boundary can be defined between harm and harmlessness of a given toxin or toxicant. This proposition is frequently used in defining limits for the emission of given pollutants to various environmental media (e.g., air, soil, water) and human populations. Many contemporary chemical pollutants used in electronics manufacturing and devices (e.g., PFAS) violate the threshold theory of harm because negative consequences are experienced on contact or in nonlinear relationships to dosages.

throughput: A measure of the energy and/or materials moving through a system, such as one devoted to manufacturing electronic devices. As energy and/or materials move through such a system pollution and waste often arise.

toxicant: A synthetic, human-made substance that is poisonous. In contrast, a toxin is a poisonous substance that originates within the living cells or organs of a biological entity such as an animal, bacteria, or plant.

transboundary flow: The movement of pollution and/or waste across jurisdictional boundaries. Such boundary crossings may be international but, can also occur at other jurisdictional scales such as across provincial, state, or municipal boundaries.

uncertainty: A situation comprised of one or more characteristics that are not fully predictable. Uncertainty describes a situation in which it is possible to know the range of states through which a system may move even if it is not possible to predict what state the system is currently in. Thus, there is a knowable range of states in which a given system may exist. In this respect, uncertainty is different from indeterminacy. Indeterminacy describes situations where the full range of states a system may be in is fundamentally unknowable or around which no fixed boundaries can be defined.

upstream: A description of the relative location of action before that which follows it and from which pollution and waste may arise. Mining of metals for electronics is upstream of manufacturing of electronics while manufacturing of electronics is upstream of use of those devices by consumers.

waste hierarchy: An analytical and interpretive device for prioritizing different waste management options. Usually the waste hierarchy is organized from the most preferred options (waste avoidance or prevention) to the least preferred (disposal).

Index

Accra, Ghana 64–5, 104, 112–13, 115, 117, 128

ACS Sustainable Chemistry & Engineering (journal) 224

Act on Recycling of Specified Home Appliances (ARSHA) 29–30

Africa 14, 25, 117
  import and export 17
  racist stereotypes 65
  waste dumping in 129

Agbogbloshie 64–5, 104, 112–15, 151, 206

Agbogbloshie Maker Space (AMS) 127–8

Akese, Grace Abena 104, 112–15

Alberta 237

Althaf, Shahana, Callie W. Babbitt, and Roger Chen, "Forecasting Electronic Waste Flows for Effective Circular Economy Planning" 221–2

aluminum scrap 29, 53–4, 77, 118–19

America 26–7
  per capita basis 14
  transboundary e-waste 18

American Chamber of Commerce in Europe (AMCHAM EU) 128–9

American Electronics Association (AEA) 235–6

André, Hampus, Maria Ljunggren Söderman, and Anders Nordelöf, "Resource and Environmental Impacts of Using Second-Hand Laptop Computers" 218–19

Apple AirPods 5, 81

Arubis (Germany) 18

Australia 27–8

Australia New Zealand Recycling Partnership (ANZRP) 28

automobile scrap 4–5

avoidance, waste 71–2, 81–6

Baldé, Cornelis Peter 72

Bamako Convention 24–5, 129, 234

Basel Action Network (BAN) 20, 130–1, 140, 162, 223, 236

Basel Convention 16, 22–5, 28, 46, 60–2, 129–32, 138, 140, 148, 151, 153–4, 163, 234

bathroom flood 44–5, 51–2

Belkhir, Lotfi, and Ahmed Elmeligi, "Assessing ICT Global Emissions Footprint: Trends to 2040 & Recommendations" 214–15

Belt and Road initiative 22

blackness and whiteness 63–4

books on pollution and waste 203–9

Bureau of International Recycling (BIR) 133

Bureau of Mines 19–20, 174, 231, 236

California 27, 199, 237

Campaign for Responsible Technology (CRT) 149, 234

Canada-US-Mexico Agreement (CUSMA) 26–7, 49–50, 68

carbon fixation 68–71, 239

cathode ray tubes (CRTs) monitors 73

Ceballos, Diana Maria, and Zhao Dong, "The Formal Electronic Recycling Industry" 222–3

cellphones 7, 13, 77

Center for Public Environmental Oversight (CPEO) 133–4

Chatham House 17, 127, 136–7

chemical toxicants. *See* toxicants

chemical vapor deposition (CVDs) tool chambers 68, 185

Chemosphere (journal) 225

China 28–9, 237
  mining and processing 22
  RoHS legislation 83

CHIPS Act 85, 134, 149, 238

chlorofluorocarbons (CFCs) 233–4

chronological list of pollution and waste 231–8

Index

Chuquicamata copper mine 1, 41, 43, 49, 174

circular economy 19, 29, 77, 106, 134, 167, 228

Circular Electronics Partnership (CEP) 122, 134–5, 144

Circular Electronics Roadmap 135

Circular Electronics System Map 135

Clean Development Mechanism (CDM) projects 186

clean production 24, 129

Clicking Clean program 146

Close the Gap 170

Code of Conduct 157–8

CO_2 emissions 54–5, 68–71, 179–82

colonialism 67, 109

Colorado 238

Commission for Environmental Cooperation (CEC) 49–50, 135–6, 235

Commodity Trade Statistics Database (UN Comtrade) 137

conflict minerals 86, 210–14

conservation 76–8, 156–8, 227, 239

copper 15–16, 18, 29, 41–4, 49, 118–19, 174–5, 212–13

Countering WEEE Illegal Trade (CWIT) 151, 186–90

Covid pandemic 111, 122, 199

cradle-to-cradle concept 123

critical minerals 22, 239

critical thinking 8–11

death date 6

Deberdt, Raphael, and Philippe Le Billon, "Conflict minerals" 209–10

decoupling 54–7, 219–20, 239

De Decker, Kris 78–9

demand-side policy 75–6

Democratic Republic of Congo (DRC) 212–13

design global, manufacture local (DGML) 92

destruction or removal efficiency (DRE) 185–6

DigitalEurope 137–8

Digital Millennium Copyright Act 79–80

disposability 5–6, 58–9, 72–4, 77, 239

documents on pollution and waste

"Center for Corporate Climate Leadership Sector Spotlight: Electronics" 182–6

"Countering WEEE Illegal Trade (CWIT) Summary Report, Market Assessment, Legal Analysis, Crime Analysis and Recommendations Roadmap" 186–90

"Fairchild, Intel, and Raytheon Sites Middlefield/Ellis/Whisman (MEW) Study Area Mountain View, California, Record of Decision" 175–9

"Flows of Selected Materials Associated with World Copper Smelting" 174–5

"Future E-waste Scenarios" 190–8

"2011–2017 Greenhouse Gas Reporting Program Industrial Profile: Electronics Manufacturing Sector" 179–82

"Nixing the Fix: An FTC Report to Congress on Repair Restrictions" 198–200

"Recovery of Precious Metals from Electronic Scrap" 173–4

Dodd-Frank Act 86, 210

do it yourself (DIY) 79, 147

downstream 2, 9, 30, 40, 43, 45, 91–3, 239

Earth system 56–7

efficiency gains 52–6, 89, 219, 239

Electronic Business (publication) 20, 236

Electronic Industry Citizenship Coalition (EICC) 157

Electronic Product Environmental Assessment tool (EPEAT) 143

Electronic Products Recycling Association (EPRA) 26, 138

electronics 2, 5–9, 239

electronic scrap 18–21, 128, 173–4, 205

electronics hibernation 122–3

Electronics Product Stewardship Canada (EPSC) 26, 138

Electronics Recycling Coordination Clearinghouse 27

Electronics Watch 138–9

246   *Index*

electronic waste (e-waste) 1, 4–6, 11
   amount and growth rates of 11–15
   Annex List 60–1, 132
   classification and definitions of 45–8
   per capita measures of 15
   regulation 21–32
   statistics 8–9
embedded software 79–80
end-of-life electronics 32, 104–5, 138, 203,
   221–2, 239–40
end-of-pipe problem 21, 24, 39–45, 127,
   129, 193, 205, 209, 223–4, 231–2,
   236, 240
Ensmenger, Nathan, "The Environmental
   History of Computing" 215
EN 45554 standard 80
environmental justice 63, 66–7, 149, 209,
   240
environmental non-government
   organizations (ENGOs) 20, 28,
   63–4, 66, 114, 130, 133, 146, 157
Environmental Pollution (journal) 225
Environmental Protection Agency (EPA)
   46–7, 70, 72, 84, 180, 186, 237, 243
environmental racism 63–4, 240
*Environmental Science and Pollution
   Research* (journal) 225–6
Environmental Science and Technology
   (journal) 226
E-Scrap News 139–40
e-Stewards 130, 140, 155, 162
European Environmental Bureau 90
European Union (EU) 22, 31–2, 58, 80, 82,
   128, 140–1, 205, 236, 238
Eurostat 140–2
"The e-Wasteland" (*Guardian* newspaper)
   108–11
Exporting Harm (report) 20–1, 130,
   223–4, 236
extended producer responsibility (EPR)
   25–7, 30–1, 74–5, 224, 240

facts and artifacts distinction 9–10
Fairphone 124
Federal Trade Commission (FTC) 165–6,
   198–200
Fitzpatrick, Colin, Elsa Olivetti, T. Reed
   Miller, Richard Roth, and Randolph

   Kirchain, "Conflict Minerals in the
   Compute Sector" 211
Fixometer 159
flat panel display manufacturers 68–9,
   181–4
flow control 22, 240
flows of e-waste 18, 22–3, 40, 43–5, 113,
   240
fluorinated greenhouse gases (F-GHGs)
   68–70, 84, 180–6
foreign garbage 28–9
forever chemicals. *See* polyfluoroalkylated
   substances (PFAS)
form factor 87, 240
Framework laptop 124
framing the e-waste problem 16–17, 21,
   32, 40, 43–5, 51, 64, 66, 71
France 80
FreeGeek 142–3

Gabrys, Jennifer, *Digital Rubbish: A
   Natural History of Electronics* 203–4
German Development Organization
   (GIZ) 114
Ghana 25, 60–1, 113–15, 206–7, 238
Gille, Zsuzsa 82
Global Electronics Council (GEC) 143
Global Enabling Sustainability Initiative
   (GESI) 144
Global E-waste Monitor (GEM) 12, 16–17,
   60, 64–5, 72
Global E-waste Partnership 161
Global E-waste Statistics Partnership
   (GESP) 14, 144–5, 150
Global North 17, 65, 103, 105–6, 112–13,
   117, 129
Global South 17, 103, 105–7, 112–13, 117
Global Witness, "The ITSCI Laundromat"
   211–12
Good Electronics Network 145–6
greenhouse gas (GHG) emissions 54–5,
   68–71, 179–83, 214–16
Greenpeace International (2008) 114, 146
Gregg, Melissa 104–5, 121–4
Grossman, Elizabeth, *High Tech Trash:
   Hidden Toxics and Human Health*
   204
Guiyu, China 28, 103, 108–10

Halvorson, Britt 63–4
Hazardous and Electronic Waste Control and Management Act 917 114
hazardous waste 22–4, 46, 49–50, 131–2, 136, 145, 240
heat transfer fluids (HTFs) emissions 181
He, Megan 9
Heppler, Jason A., "Green Dreams, Toxic Legacies: Toward a Digital Political Ecology of Silicon Valley" 215–16
Hewlett-Packard (HP) 20
Hieronymi, Klaus, Ramzy Kahhat, and Eric Williams, *E-Waste Management: From Waste to Resource* 204–5

IBM facility
    cancer cluster at 233
    chemical toxicants at 235
iFixit 147
incommensurability 15, 47–8, 240
indeterminacy 46–8, 50, 240, 243
industrial manufacturing capacity 5–6, 22
informal electronic waste 117
informal e-waste sector 114–15
Information Technology Industry Council (ITI) 147–8
Institute of Scrap Recycling Industries (ISRI) 154
International Campaign for Responsible Technology (ICRT) 148–9, 234
International Criminal Police Organization (Interpol) 150–1
International Merchandise Trade Statistics (IMTS) 137
International SEMATECH Manufacturing Initiative 181–2
International Solid Waste Association (ISWA) 149–50
International Telecommunications Union (ITU) 150
International Tin Supply Chain Initiative (ITSCI) 211–12
Inventory of Existing Chemical Substances Produced or Imported in China (IECSC) 84
Isenberg, Nancy 64
IT asset disposition 122

Japan 29–30, 238
Jevons Paradox. *See* rebound effect
Jevons, William Stanley 52–5, 242
Journal of Cleaner Production (journal) 226
Journal of Hazardous Materials (journal) 226–7
"Just Say No to E-waste" (report) 236

Kahhat, Ramzy 103, 105–7
Kaitatzi-Whitlock, Sophia 208
Kuehr, Ruediger, and Eric Williams, *Computers and the Environment* 205

language learning models (LLMs) 56
Law for the Promotion of Effective Utilization of Resources (LPEUR) 29–30
leachate 41, 241
Lécuyer, Christophe, "From Clean Rooms to Dirty Water" 216
legality/illegality 60–2
Lepawsky, Josh, *Reassembling Rubbish: Worlding Electronic Waste* 206
Liboiron, Max 67
lifecycle assessment (LCA) 218–19
lifecycle of electronics 67, 83, 127, 139, 148–9, 163, 204–5, 209, 241
lifetime average daily dose (LADD) 47
lightweighting 56, 241
Linear Growth scenario, e-products consumption 191–3
liquid crystal display (LCD) panels 182–4
Little, Peter C.
    Burning Matters 206
    Critical Zones of Technopower and Global Political Ecology 207
    Toxic Town: IBM, Pollution, and Industrial Risks 206

MacBride, Samantha 2, 76
magnitude of e-waste 14–15, 43–5, 48–50, 241
Magnuson-Moss Warranty Act (MMWA) 165, 198, 200
Masanet, Eric, Arman Shehabi, Nuoa Lei, Sarah Smith, and Jonathan

248 *Index*

Koomey "Recalibrating Global Data Center Energy-Use Estimates" 219–20

massively open online course (MOOC) 169

material flow analysis (MFA) 221–2

material heterogeneity 87

Maxwell, Richard, and Toby Miller, *Greening the Media* 207

Maxwell, Richard, Jon Raundalen, and Nina Lager Vestberg, *Media and the Ecological Crisis* 207–8

Melosi, Martin 71

Middle East/West Asia 25

Middlefield/Ellis/Whisman (MEW) Study Area 175–9, 237

million metric tons of carbon dioxide equivalent (MMT CO_2e) 179–80

mining process 1, 9, 21–2, 39–41, 43, 72–3, 75, 93, 209–14

Minter, Adam 29, 103, 108–11

modern waste 2–8, 241

modularity 87–8, 91

mopping up 45, 52

municipal solid waste (MSW) 8–9

National Center for Electronics Recycling (NCER) 27

National Environmental Security Task Forces (NESTs) 151

National Strategy for Electronics Stewardship (NSES) 27

National Television and Computer Recycling Scheme (NTRS) 28

National Voluntary Partnership for E-waste Recycling (NVPER) 30

New York 238

Nigeria 25, 66, 237

no export approach 59

non-governmental organizations (NGOs) 90, 167, 170, 241

North American Free Trade Agreement (NAFTA) 26, 49, 135–6, 235

Northeast Recycling Council (NERC) 27

Ntapanta, Samwel Moses, *Mahakama ya friji* (Refrigerator Court) 104, 117–20

O'Brien, Mary 88

occupational and environmental health risks 223

occupational health and safety (OHS) 139

Oeko-Institut 151–2

O'Neill, Kate, *Waste* 208

Open Repair Alliance 152

Oregon state 80, 142

Organisation for European Economic Co-operation (OEEC) 153

Organization for Economic Co-operation and Development (OECD) 22, 153

original equipment manufacturers (OEMs) 79–81, 122, 124, 198

Out of Box Experience (OOB) 121–2

Pacific Studies Center (PSC) 133–4, 232

Parajuly, Keshav, James Green, Jessika Richter, Michael Johnson, Jana Rückschloss, Jef Peeters, Ruediger Kuehr, and Colin Fitzpatrick, "Product Repair in a Circular Economy" 220–1

parts pairing 79–80, 88, 157

perfluorocarbon (PFC) emissions 181–2

personal computer 29, 124, 142, 205

planned obsolescence 5–6, 241

plastics 2, 77

polluter-pays-principle 25, 74–5, 241

pollution and waste 1, 7–9, 21, 24–5, 30–3, 39–41, 45–55, 67–8, 70–5, 77–8, 81–6, 89, 93, 127, 135–6, 145–9, 158–64, 173, 204–7, 209–10, 213–19, 225–7, 241

pollution release and transfer registries (PRTRs) 49, 85–6, 135–6, 241

polychlorinated biphenyls (PCBs) 23, 45

polyfluoroalkylated substances (PFAS) 68, 70–1, 84–5, 88, 134, 238

post-consumer e-waste 1, 12–15, 21, 30–3, 43, 48–52, 57, 59, 67, 71, 75–6, 82–3, 137–8, 143, 145–9, 158, 163, 168, 170, 186, 221–4, 241

precautionary principle 24, 242

preservation 76–7, 242

Proactive Approach, e-products consumption 195–7

Index

producer responsibility organizations (PROs) 138, 166–7

Project Eden 151

Promotion of Recycling of Small Waste Electrical and Electronic Equipment (PRSWEEE) Act 30

Puckett, Jim, Leslie Byster, Sarah Westervelt, Richard Gutierrez, Sheia Davis, Asma Hussain, and Madhumitta Dutta, "Exporting Harm-The High-Tech Trashing of Asia" 223–4

race and waste 63–4

Reactive Approach, e-products consumption 193–5

"Ready for Recycling?" (article) 236

reBOOT 153

rebound effect 53–6, 78, 88–9, 182, 220, 242

"Recovery of Precious Metal from Electronic Scrap" (report) 231

Recycled Materials Association (ReMA) 154

Recycling and Waste Reduction Act (RWRA) 28

recycling, concept of 74–8, 242

Recycling Industry Operating Standard (RIOS) 154–5

recycling loads 41

Reduction of Hazardous Substances (RoHS) 31–2, 83–4

reduction, waste 81–6

refrigerator 104, 117–20

Registration, Evaluation, Authorisation and Restriction of Chemicals (REACH) regulations 31–2, 84

regulation of e-waste 21–2. *See also individual countries*

Reno, Joshua O. 63–4

repair and maintenance (R&M) 106

Repair Association 79–80, 155

repair café movement 156

Repair Europe 156–7

resource on pollution and waste 19–20, 53, 242

books 203–9

journals 224–8

manufacturing 214–18

mining 209–14

post-consumer discard 221–4

use and reuse 218–21

Resource Recycling of Electrical and Electronic Equipment and Vehicles (RREEV) Act 31

Resource Recycling Promotion Law (1991) 29

Resources, Conservation, and Recycling (journal) 227

Responsible Business Alliance (RBA) 157–8

Responsible Environment Initiative (REI) 158

Restart Project 158–9

reuse 78–81

reuser experience 121–2

reverse logistics 122

right to repair 79–81, 93, 147, 154–7, 165–6, 198–9, 238

Roberts, Hedda 91

Roy, Ananya 65–6

Santa Clara Center for Occupational Safety and Health (SCCOSH) 148–9, 159–60, 231

scalar mismatch 51–3, 93, 211, 213, 242

Science and Technology Studies (STS) 9

Science of the Total Environment (journal) 227

scrap 4–5, 18–21, 128, 173–4, 205

scrapyard 110, 113–15, 128

secondary lead processing 18–19

Second World War 2, 4–5, 153

SEMATECH 234–5

serialization 88

Siegel, Lenny, and John Markoff, *The High Cost of High Tech* 208–9

Silicon Valley 46, 67, 73–4, 82, 86, 133–4, 159, 175, 204, 209, 215–16

"Silicon Valley: Paradise or Paradox?" (report) 232

Silicon Valley Toxics Coalition (SVTC) 82–3, 149, 159–60, 232–4, 236

Singapore 30–1

slumdog urbanism 65

Smith, Maureen 76

Smith, Ted, David Allan Sonnenfeld, and David N. Pellow, *Challenging the Chip* 209
solutions
  design 86–8
  reorganizing for 89–93
Solve the E-waste Problem (StEP) 160–1
South Korea 31, 48, 237
stock 45, 90–2, 242
Strasser, Susan 4, 6
sufficiency 56, 89–90, 242
superfund 46–7, 73, 83, 232, 243
supply-side policy 75–6
Sustainable Cycles program 164
Sustainable Electronics Recycling International (SERI) 161–2

3TG minerals 213, 210–11
threshold theory 45, 74, 136, 141, 223, 243
throughput, concept of 40, 53–6, 78, 89, 221–2, 243
toxicants 32, 44, 49–50, 52, 68, 72–3, 82–5, 88, 93, 120, 133, 139, 159–60, 176, 209, 216, 223, 231–3, 235, 243
Toxic Release Inventory (TRI) 233
Toxic Substances Control Act (TSCA) 84
transboundary flow/shipments 15–19, 22–4, 29, 43, 57–61, 131, 137, 141, 205, 243
trichloroethylene (TCE), manufacturing electronics 46–7, 73, 233

Umicore (Belgium) 18
uncertainty 8–9, 46–7, 50, 197, 243
United Nations Conference on Trade and Development (UNCTAD) 162
United Nations Environment Programme (UNEP) 163
United Nations Industrial Development Organization (UNIDO) 163–4
United Nations Institute for Training and Research (UNITAR) 164
United Nations University (UNU) 164–5
United Nations University/United Nations Environment Programme (2019) 190–8
United States Department of the Interior (1972) 173–4

United States Environmental Protection Agency (1989) 175–9
United States Environmental Protection Agency (2016) 182–6
United States Environmental Protection Agency (2018) 179–82
United States Federal Trade Commission (2021) 165–6, 198–200
United States Geological Survey (USGS) 41, 174–5
United States Geological Survey, and Thomas G. Goonan, "Flows of Selected Materials Associated with World Copper Smelting" 212–13
United States Public Interest Research Group (US PIRG) 166
upstream 9, 25, 30, 40, 43, 45, 51, 68, 91–2, 243
urban mines 73
The US Federal Remediation Technologies Roundtable 237

Vogel, Christoph, and Timothy Raeymaekers, "Terr(it)or(ies) of Peace? The Congolese Mining Frontier and the Fight against 'Conflict Minerals'" 213–14

waste, concept 3–4, 19–20
waste dumping 57–9, 64, 66, 103, 113, 129, 206
waste electrical and electronic equipment (WEEE) 31–2, 141, 149, 166–7, 236
waste hierarchy 71–2, 243
  design solutions 86–8
  disposal 72–4
  recycling 74–8
  reduction, avoidance and prevention 81–6
  reorganizing for solutions 89–93
  reuse 78–81
Waste Management (journal) 227–8
Waste Management & Research: The Journal for a Sustainable Circular Economy (journal) 228
WEEELABEX 167–8
white savior industrial complex (WSIC) 66

Index

Williams, Eric, R. U. Ayres, and M. Heller, "The 1.7 Kilogram Microchip" 217
World Computer Exchange (WCE) 168
World Environment Situation Room 163
World Health Organization (WHO) 168–9
World LCD Industry Cooperation Committee (WLICC) 184
WorldLoop 170

World Reuse, Repair, and Recycling Association (WR3A) 169–70
Wynne, Brian 46

Yu, Jinglei, Eric Williams, and Meiting Ju, "Analysis of Material and Energy Consumption of Mobile Phones in China" 217–18

Zimmerman, Erich W. 19–20

About the Author

Josh Lepawsky is Professor of Geography at Memorial University of Newfoundland and Labrador. He is fascinated by connections between geography, technological systems, and their discards. Questions informing his research include: where and how are contemporary discards made? Where do they travel and where do their effects accumulate? Who gets what discards, where, how, and under what conditions? He is also interested in how maintenance and repair, broadly conceived, might offer both literal and figurative lessons for figuring out how to live well together in permanently polluted and always-breaking worlds. More about his work can be found at https://electronicplanet. xyz/about/blog/.